The Remaking of the Mining Industry

The Remaking of the Mining Industry

David Humphreys

palgrave
macmillan

First published 2015 by
PALGRAVE MACMILLAN

Palgrave Macmillan in the UK is an imprint of Macmillan Publishers Limited, registered in England, company number 785998, of Houndmills, Basingstoke, Hampshire RG21 6XS.

Palgrave Macmillan in the US is a division of St Martin's Press LLC, 175 Fifth Avenue, New York, NY 10010.

Palgrave Macmillan is the global academic imprint of the above companies and has companies and representatives throughout the world.

ISBN 978–1–137–44200–0

This book is printed on paper suitable for recycling and made from fully managed and sustained forest sources. Logging, pulping and manufacturing processes are expected to conform to the environmental regulations of the country of origin.

A catalogue record for this book is available from the British Library.

Library of Congress Cataloging-in-Publication Data
Humphreys, David, 1952–
The remaking of the mining industry / David Humphreys.
pages cm
Summary: "The emergence of China as a major economic power in the first decade of the millennium prompted the biggest commodity boom of modern times. Soaring prices gave rise to talk of a commodity 'super cycle' and induced a wave of resource nationalism in mineral-rich countries. It also stirred up concerns of supply shortages in mineral-consuming countries. The author, who served as chief economist at two of the world's largest mining companies during these years, describes how and why this resulted in a transformation – a 'remaking' – of the mining industry. The book tells of how the markets in which the industry operated changed, how the industry was restructured through acquisition and investment, and how a cast of new players from emerging economies arrived on the scene. With the boom now passed, the book concludes with some reflections on what the changes imply for the future of the industry and the environmental and political challenges it will face" — Provided by publisher.
ISBN 978–1–137–44200–0 (hardback)
1. Mineral industries—China. 2. Economic development—China. I. Title.
HD9506.C62H86 2015
338.20951—dc23 2015003230

Contents

List of Figures and Tables vi
Acknowledgements viii
About the Author ix
List of Abbreviations x

Introduction 1
1 The Barren Years 12
2 China Changes Everything 38
3 Feeding the Dragon 61
4 Prices Take Wing 88
5 An Industry Transformed 114
6 Enter the Emergers 140
7 Dividing the Spoils 159
8 Consumer Concerns 185
9 The Next Thirty Years 206

Notes 232
Selected Bibliography and Data Sources 245
Index 250

Figures and Tables

Figures

Cover Image: The mine featured is Rio Tinto Kennecott's Bingham Canyon mine near Salt Lake City, Utah, USA. In the background are the snow-covered peaks of the Oquirrh Mountain Range. Bingham Canyon is one of the world's largest man-made excavations. It is 2½ miles across and ¾ mile deep and is visible from space. © Rio Tinto

1.1	Intensity of use of selected metals, 1950–2000	14
1.2	Metal prices, 1978–2002	21
1.3	Profitability by industry sector, 1973–1999	23
2.1	Growth in China's exports and industrial production, 2000–2014	42
2.2	China: Composition of GDP, 1980–2013	45
2.3	Use of metals in China	49
2.4	Composition of global growth, 1985–2013	51
3.1	World imports of seaborne iron ore, 1978–2013	65
3.2	China's overseas direct investment in mining, 2003–2012	82
4.1	Prices of selected mineral commodities, 2000–2014	93
4.2	Copper cost curves, 2004 and 2007	96
4.3	Prices of internationally traded products, 1990–2014	100
4.4	US trade-weighted dollar index, 1975–2014	101
4.5	Price of seaborne iron ore, 2000–2014	111
5.1	Mineral exploration expenditure by commodity, 1994–2014	131
5.2	Mineral exploration expenditure by company type, 1997–2014	132
5.3	Mining investment and prices, 1978–2013	134
5.4	Return on capital employed: 40 largest mining companies	135
6.1	Composition of top 40 miners, 2003–2013	142
6.2	Global distribution of mineral reserves	148
7.1	Mineral exploration expenditure by region, 2000–2013	162
7.2	Polarisation of global mineral supply	176
8.1	Composition of global GDP and exports of goods and services, 2013	186

8.2 World production of rare earths, 1960–2013 198
9.1 Distribution of global mineral consumption, 2000 and 2013 210

Tables

I.1 Structure of the mining industry, 2012 10
2.1 Growth in consumption of selected commodities: China and the world 47
3.1 China's imports of selected mineral raw materials, 2000–2013 64
3.2 China's mine production, 2000 and 2013 72
5.1 World's largest mining companies by value of mine production of metals, 2000 118
5.2 World's largest mining companies by value of mine production of metals, 2008 124
7.1 World mine production of selected mineral commodities in 2013 160
7.2 Ranking of countries according to economic contribution of mining 164
7.3 Expenditure by members of the World Gold Council, in $ million 182
8.1 Mine production in traditional mineral-consuming countries, 1990 and 2013 189

Acknowledgements

This book is the product of a lifetime spent in the mining industry and in working on mining-related policy. Accordingly, it owes much to many people spread over a long period of time. None of these unwitting contributors is, of course, implicated in any way with the views expressed in the book or any errors of fact, the responsibility for which lies squarely with the author.

Among the many who have wittingly or otherwise contributed to the thinking contained in this book, I would like to mention particularly my former colleagues at Rio Tinto, Phillip Crowson and Michael Bailey, along with my friends John Tilton, Marian Radetzki, Gustavo Lagos and Rod Eggert, with whom I have passed many hours in congenial discussion about mining economics. I thank Marian Radetzki, John Tilton, Fiona Perrott-Humphrey and Magnus Ericsson for reading and commenting on drafts of various chapters. I am also grateful to David Wilson, Olga Antimonova, Michael Komesaroff and Ken Haddow for their assistance during the preparation of this book. Finally, I thank my family, who over the years had to put up with many absences as I plied my trade in this most global of industries and also for their helpful suggestion that this book be titled 'Mine Life'.

The author and publisher are grateful to the following for permission to use copyright materials: Elsevier, Macquarie Capital, Rio Tinto, SNL Finance, Urandaline Investments and Wood Mackenzie.

About the Author

David Humphreys was formerly chief economist of the London-based mining company, Rio Tinto, and of Russia's largest mining company, Norilsk Nickel. He also served on the board of Russian gold miner Petropavlovsk. Prior to entering the mining industry, David worked in UK government service as an advisor on minerals policy. He was co-founder and first vice president of the industry federation Euromines. He has been a visiting scholar at the Colorado School of Mines and the Catholic University of Chile in Santiago, has been a Bosch Fellow at the Transatlantic Academy in Washington DC and is an honorary lecturer at the University of Dundee. He is an independent consultant and writes a regular column for *Mining Journal*. He has a bachelor's degree and a PhD from the University of Wales.

Abbreviations

AIG	Australian Institute of Geoscientists
AIM	Alternative Investment Market (of the LSE)
AMV	Africa Mining Vision
ANC	African Mining Congress
ASX	Australian Securities Exchange
AusIMM	Australasian Institute of Mining and Metallurgy
BEE	Black Economic Empowerment (South Africa)
BRIC	Brazil, Russia, India and China
CCP	Chinese Communist Party
CEO	Chief Executive Officer
CFTC	Commodity Futures Trading Commission (US)
CFR	Carriage and Freight
CIC	China Investment Corporation
CISA	China Iron and Steel Association
CITIC	China International Trust and Investment Corporation
CMI	Critical Minerals Institute (US)
CNMC	China Nonferrous Metal Mining (Group) Company
CRM	China Railways Materials Corporation
CTA	Commodity Trading Advisor
CVRD	Companhia Vale do Rio Doce
DoD	Department of Defense (US)
DOE	Department of Energy (US)
DRC	Democratic Republic of the Congo
EAF	Electric Arc Furnace
EC	European Commission
EIA	Energy Information Administration (US)
EITI	Extractive Industries Transparency Initiative
ENRC	Eurasian Natural Resources Corporation
ETF	Exchange Traded Fund
EU	European Union
FAI	Fixed Asset Investment
FCPA	Foreign Corrupt Practices Act (US)
FDI	Foreign Direct Investment
FIRB	Foreign Investment Review Board (Australia)

FOB	Free on Board
FQM	First Quantum Minerals
FSU	Former Soviet Union
GATT	General Agreement on Tariffs and Trade
GDM	Global Diversified Miner
GDP	Gross Domestic Product
GFC	Global Financial Crisis
GFCF	Gross Fixed Capital Formation
GFMS	Gold Fields Mineral Services
GMI	Global Mining Initiative
HKEx	Hong Kong Stock Exchanges and Clearing Ltd
ICMM	International Council on Mining and Metals
ICSG	International Copper Study Group
ICT	Information and Communications Technology
IEA	Institute of Economic Analysis
IFC	International Finance Corporation
IIED	International Institute for Environment and Development
IMF	International Monetary Fund
IP	Industrial Production
JOGMEC	Japan Oil, Gas and Metals National Corporation
JORC	Joint Ore Reserves Committee (Australia)
LBMA	London Bullion Market Association
LME	London Metal Exchange
LPPM	London Platinum and Palladium Market
LSE	London Stock Exchange
M&A	Mergers and Acquisitions
MCA	Minerals Council of Australia
METI	Ministry of Economy, Trade and Industry (Japan)
METS	Mining Equipment, Technology and Services
MIIT	Ministry of Industry and Information Technology (China)
MITI	Ministry of International Trade and Industry (Japan)
MINT	Mexico, Indonesia, Nigeria and Turkey
MMG	Minerals and Metals Group (China)
MMSD	Mining, Metals and Sustainable Development
MoU	Memorandum of Understanding
MPRDA	Mineral and Petroleum Resources Development (South Africa)
MRRT	Minerals Resource Rent Tax (Australia)
NBS	National Bureau of Statistics (China)

NDRC	National Development and Reform Commission (China)
NGO	Non-Governmental Organisation
NI43-101	National Instrument 43-101 (Canada)
NRC	National Research Council
OECD	Organisation for Economic Co-operation and Development
OPEC	Organization of the Petroleum Exporting Countries
PCS	Potash Corporation of Saskatchewan
PDAC	Prospectors and Developers Association of Canada
PGM	Platinum Group Metal
PNG	Papua New Guinea
PPP	Purchasing Power Parity
PwC	PricewaterhouseCoopers
RICI	Rogers International Commodity Index
RKEF	Rotary Kiln Electric Furnace
RMB	Renminbi
RMD	Raw Materials Database
ROCE	Return on Capital Employed
SAFE	State Administration of Foreign Exchange (China)
SASAC	State-Owned Assets Supervision and Administration Commission (China)
SEC	Securities and Exchange Commission (US)
SOE	State-Owned Enterprise
SRB	State Reserve Bureau (China)
SXEW	Solvent Extraction Electro Winning
TCRC	Treatment Charge and Refining Charge
TSX	Toronto Stock Exchange
UNCTAD	United Nations Conference on Trade and Development
US	United States of America
USBM	US Bureau of Mines
USGS	US Geological Survey
VLOC	Very Large Ore Carrier
WBCSD	World Business Council for Sustainable Development
WBMS	World Bureau of Metal Statistics
WCA	World Coal Association
WGC	World Gold Council
WSA	World Steel Association
WTO	World Trade Organization

Introduction

The first decade of the twenty-first century saw the biggest and most sustained commodity boom of modern times, fuelled by China's rapid industrialisation.

The rise in mineral commodity prices transformed the fortunes of the mining industry. An industry which had become reconciled to its fate as an unprofitable slow-flowing backwater suddenly became the object of widespread investor enthusiasm. There was talk of a commodity 'super cycle'. Companies in the sector started to dust off old projects in order to expand their production and capture the opportunities which the boom was generating. Some sought to shortcut the process of growth by buying other companies. For the first time in many years, companies looked to take on new staff. Exploration companies started popping up all over, and miners from the emerging market economies began to assert themselves on the world stage. Governments of mineral-rich countries, which had earlier become disenchanted with the ability of mining to promote economic development, began to look on the sector with fresh eyes as revenues resulting from higher commodity prices piled up. At the same time, governments of the traditional mineral-consuming countries started to fret that shortages of key mineral commodities and high prices might constrain growth in key economic sectors and compromise national security.

Ten years on and the boom has passed. The tidal wave of investment unleashed by the boom is feeding through into increased levels of mineral production. Markets have moved back into better balance or into surplus. China's economy continues to expand, but its major growth surge is behind it. Lower prices have sharply reduced industry margins and forced the write-down of acquisitions made when spirits were running high. As companies cut their costs and seek to repair their balance sheets, the language they are employing about the importance of capital management is reminiscent of that used in the years before the boom. Increasingly, it looks as though the boom was just another cycle, albeit a particularly strong and sustained one.

However, if the market conditions facing the mining industry have come full circle, the industry is a very different place from what it was. The boom changed it in some important and lasting ways. The structure of the industry was transformed by the wave of mergers and acquisitions which accompanied booming prices. Many middle-ranking mining companies disappeared. Mining companies from the emerging world, which rose to prominence in the boom years, have become permanent features of the industry landscape. There are a whole host of new investors in the sector, including a range of state-backed companies, sovereign wealth funds, oligarchs and private equity investors. The customer base of the industry has shifted towards Asia, a shift that has resulted in changes both to the forms in which minerals are traded and how they are priced. Mineral commodity prices, while they have fallen back from the peaks they attained at the height of the boom, have not fallen back to where they started out before the boom. The long downward trend in real mineral prices experienced in the quarter century following the boom of the 1970s appears to have been broken.

As a result of changing attitudes in mineral-host countries and the poor public perception of mining going into the boom, companies within the industry have become more sensitised to the impacts of their activities as well as to the potential part they can play in promoting economic development. The geopolitical context in which the industry has to operate in the post-boom years is very different from that which existed before. Since the global financial crisis (GFC) of 2008, the balance of global economic and political power has shifted towards China and to other emerging market countries. Nationalist sentiment has emerged as a strong driver of economic and industrial policy, and in many parts of the world the state is becoming more closely engaged in the organisation and direction of the minerals sector.

The boom also served to transform awareness about the industry and attitudes towards it. For much of the time, mining is obliged to live in the shadow of the much larger and geopolitically important oil and gas industries. Because of the wide range of minerals and mineral producers, the mining industry is highly fragmented. Outside the major mining countries, it rarely makes headlines. Since the minerals that people use are generally incorporated into complex consumer products such as cars or phones, the public doesn't have direct exposure to minerals in the same way they have when they go to the pump to fill their cars. To the extent that mining impinges on peoples' consciousness at all, it is often when something goes wrong, when miners get trapped underground, when the revenues of mining are found to be funding rebel activities or when there is some sort of environmental disaster.

The boom helped put mining back on the map and prompted a re-evaluation of its role. The shift in the focus of mineral demand away from the slow-growing de-industrialising West towards fast-growing populous emerging economies gave mining renewed economic and political relevance. It forced mining company managers to think about their industry in more expansive terms than they had become used to. It encouraged investors to think of mining as a profitable growth sector alongside sectors such as IT and biotech rather than as the stolid margin-based industry they had seen it as before. London was able to re-invent itself as a global centre of mining finance. Young people who in an earlier era would not have contemplated mining began to view it as a potentially exciting and lucrative career option. Emerging economy countries, which for years had grappled with the notion that the possession of natural resources could be a 'curse', began to think more constructively about how to use their mineral wealth to benefit their citizens. And consumers the world over were forcibly reminded about the critical role that minerals play in supporting and improving their standards of living, whether in the form of basic raw materials used to build houses and roads or the high-technology metals required to make smart phones. Mining, while never likely to be the object of public affection, was at least earning some new-found respect.

This book is about the transformation – the 'remaking' – of the mining industry, which occurred during these years. It seeks to explain how the boom came about, why it came about and what its consequences were. And it offers some reflections on the significance of the changes that took place and what they might mean for the future of the industry.

The book begins, in Chapter 1, with a review of conditions in the mining industry in the years leading up to the commodity boom. The 1980s and 1990s were generally grim years for mining. The prolonged economic recession which followed the 1970s oil shocks and the collapse of the Soviet Union had suppressed global demand for minerals. There was too much mine capacity, and mineral prices were persistently weak. By the end of the 1990s, investors had largely lost interest in mining. Their preoccupation was technology and the post-industrial 'weightless' economy. Battered by years of poor profitability and seemly faced with a long-term decline in real (inflated-adjusted) commodity prices, the mining industry had little interest in investment for growth. Its focus was on reducing costs, increasing capital efficiency and improving its social and environmental performance so as to counter negative public and investor perceptions of the industry.

Chapter 2 relates how, during the course of the first decade of the new millennium, China dramatically transformed the prospects of the mining industry. Although the reforms that led to this were laid down years before, China's economy had up to this point lacked the critical mass to have much impact on the global economy or the global demand for minerals. China's joining the World Trade Organization (WTO) in 2001 and the Zhu Rongji reforms which led up to this, triggered a massive boom in exports as China established itself as the industrial workshop of the world. The wave of investment which underpinned and sustained this development, and the rapid urbanisation which accompanied it, was highly material intensive, and China's demand for mineral commodities of all types soared.

The rapid growth in China's demand for minerals prompted a corresponding need for rapidly increased supply. This is the subject of Chapter 3. Although China is a major mineral producer, it is not rich in the resources of certain important industrial raw materials such as iron ore, bauxite, copper and nickel. Accordingly it began to import these minerals in increasing quantities. It also encouraged rapid growth in domestic Chinese production, often at significant environmental cost, and at the same time discouraged the export of locally produced mineral products. From 2004 onwards, China sought to boost mineral supply by launching a programme of overseas investment in mining spearheaded by state-backed companies, development organisations and its sovereign wealth fund. Particular targets for this investment were Australia and countries in Africa.

Chapter 4 looks at the impact of these developments on mineral prices. Mineral producers, after years of underinvestment, were unable to respond to the upsurge in Chinese demand. The scale and persistence of the resulting price increases had by 2005 given rise to the notion that commodity markets were embarked on a multi-decade 'super cycle' driven by expanding demand from China and other emerging market countries. There were also suggestions that a rise in investor interest in commodities was serving to exaggerate price movements. Both ideas are investigated and found wanting. However, there do appear to be grounds, rooted in rising marginal costs of production, shifting terms of trade and in long-term currency developments, for supposing that the

30-year downtrend in real dollar commodity prices was broken by the boom. It is also apparent that China had an important effect on price dynamics, prompting a move in the pricing of bulk commodities such as iron ore from a long-term contract to a spot basis.

The changed price environment fundamentally transformed the prospects of the mining industry. Chapter 5 records how, following years of neglect, the industry was suffering shortages of projects and skilled personnel and was slow to respond to the new situation. Given how long it takes to get new mines into production, some companies decided to take a shortcut to volume growth, and 2005–2007 saw a wave of mergers and acquisitions (M&A) in the industry, many of these involving the large diversified mining companies. The financial markets fell in love with the sector once more, bankrolling corporate growth ambitions and funnelling money towards exploration companies. Although knocked back by the GFC, the industry quickly recovered thanks largely to China's massive stimulus programme and investment in the sector continued to rise through to 2011. Expansion of the industry's production capacity coupled with a slowing of growth in China subsequently brought markets into better balance and saw prices soften. By 2012, companies were beginning to prune their capital expenditure and to direct their efforts towards boosting margins and returns to shareholders.

The other major corporate changes brought about by the commodity boom were among mining companies from emerging market countries. These changes are the subject of Chapter 6. The wave of corporate privatisations which occurred in the 1990s, combined with the opening up of capital markets, enabled many of these companies to grow strongly through the boom years and to begin investing outside their traditional areas of operation. Mining companies from Russia, Kazakhstan, India, Peru and Mexico, among others, listed on the London Stock Exchange. State-backed mining companies from China also became prominent on the world stage. The risk appetite of these companies and their often pragmatic approach to business appeared to make them well adapted to the new conditions facing the industry and potentially fierce competition for the established global miners. While weaker market conditions have revealed several problems with their businesses models and with standards of corporate governance, they are set to play an important part in the mining industry of the future.

Chapter 7 examines how the commodity boom influenced the attitudes and behaviour of mineral-rich countries and how the benefits of the boom were distributed. Rising prices naturally boosted the fortunes of mineral-producing countries from Australia to Africa and led to them attracting large inflows of capital. In principle, the boom represented a good development opportunity for these countries, but in practice managing mineral-driven economies can be challenging. Attempts by mineral-rich countries to secure the full benefit from the boom for their economies led to higher taxes on mining, restrictions on foreign ownership of mining assets and export controls, a tendency commonly referred to as resource nationalism. It is suggested that there are structural factors that make it unlikely this will dissipate quickly. Although the large mining companies have devoted considerable efforts to trying to be good corporate citizens and to understanding how their investments can be better used to leverage development in mineral host countries, these efforts have had to a significant degree been negated by the growing politicisation of mining and mining regimes.

The same factors which were making mining a rewarding activity for mineral-rich countries were giving rise to concerns over the security of supplies in mineral importing and consuming countries. Chapter 8 considers the responses and policy measures adopted by this group of countries, focusing on the United States and the European Union (EU). Particular attention in these countries was directed towards a small group of high-tech minerals produced in China, including rare earths, indium, gallium and germanium. Policy proposals for dealing with supply threats to these minerals focused on improving the efficiency of resource use, researching possible substitutes and ensuring that trade in mineral products was free and fair. However, there appeared little appetite for encouraging more domestic mining in the United States and EU as part of the policy response. In fact, the share of these areas in global mining had been on a declining trend for many years and the interest of governments in mining has been largely confined to issues of governance and transparency among mining companies listed on their stock exchanges. In protest at what they deemed to be unfair restrictions on the export of mineral products, the United States and the EU brought two cases against China in the WTO, the second of these relating to rare earths. It is argued that the preoccupation with a small group of exotic minerals from one country distracted policy-makers in the United States and Europe from the broader systemic challenge to the supply of all mineral materials posed by the rise of resource nationalism.

Chapter 9 considers where the influences of the past few years are trending. The surge in demand for minerals associated with China's rapid industrialisation has passed, but industrial growth in the emerging world and the urbanisation which will accompany this will underpin growth in the demand for mineral products for years to come, probably at rates in line with historical experience. Changes in the customer base of the industry have brought with them increased reliance on spot pricing in mineral markets, and this will give rise to greater price volatility and increased challenges for mining investment. Terms of trade effects and cost pressures are likely to sustain prices (although not margins) at levels above those prevailing before the commodities boom. Physical resource availability should not constrain future mineral developments, but there are two major 'above ground' factors which have the potential to do so. These are increased pressures on mining and mining costs arising from environmental impacts (on land, water and air) and political uncertainty. The retreat from globalisation is favouring national and statist interests within many emerging market countries. Global miners are adapting to this situation by restructuring their businesses and refocusing themselves geographically.

Industry histories are inevitably products of their time. They seek to impose a shape on a lot of disparate facts and events, some of which are still too recent for the observer to get a full and balanced perspective on them. Some things that seem important now will doubtless seem rather trivial to readers in ten years' time, while other events which are touched on only in passing will turn out to be more important than currently supposed.

There are many books on mining but not many on the mining industry. One of the few that exists was written by a former US mining executive, Simon Strauss, in the mid-1980s. The book was entitled, rather enigmatically, *Trouble in the Third Kingdom.*[1] The 'Third Kingdom' is a reference to the last element in the taxonomy of 'animal, vegetable and mineral' established by the Swedish botanist Carl Linnaeus. In this book, Strauss chronicled the achievements of the mining industry in the post–World War II period through to the 1970s before its fall from grace in the early 1980s (the 'trouble' to which he refers). The present book effectively picks the story up where Simon Strauss left it. It tells how the industry moved on after the ending of the Cold War and the long period of depressed commodity markets before being propelled back into the lime light during the present century.

The coverage and perspectives of this book inevitably reflect the knowledge and experience of the author. So, it is perhaps best that these be made explicit at the outset. For much of the period covered by the book, I was chief economist at two of the world's largest mining companies, one based in an advanced economy and one in an emerging market country. I have subsequently consulted for a range of mining companies and international organisations and served as a non-executive director with a UK-listed gold company. These posts afforded me a ringside view of many of the events described in the book. Along the way, I sought to figure out what I was observing and to communicate my thoughts through various publications and presentations, some of which I have drawn on in writing this book.

As a consequence of my background, the book has a pronounced 'large company' perspective. This is not intended to be disrespectful to the small company sector where many of the industry's most exciting and creative developments occur, but it is what I know. The same applies to the book's geographical focus. Although the book seeks a global focus for what is, after all, a global industry, it may come across to readers in some parts of the world as unduly London-centric. Again, this is unfortunate since much of interest in the industry is going on far from London, although I would still argue that there are few better places in the world than London – a truly global city – to seek to build an understanding of a global industry.

Finally, I should set out the parameters of the book. Mining as discussed here is the industry devoted to the recovery of hard minerals from the ground. It thus includes the mining of base metals such as copper, precious metals such as gold, gem stones such as diamonds, non-metallic minerals such as phosphates, and coal. It excludes oil and gas and also large-volume low-value mineral materials used for construction. Many of the illustrations in the book are drawn from among the metals that are, along with coal, the most valuable mined products and the best documented. Judged by value of metal in the ore, the three most economically important metallic minerals are iron ore, gold and copper. Although the relative value of these minerals fluctuates year to year along with prices, these three metallic minerals are estimated to account for around two-thirds of the total value of metals mined.[2] Other important metallic minerals include nickel, lead, zinc, tin, silver and the platinum group metals (PGMs).

The term 'mineral' perhaps also requires a word of explanation. For the geologist, 'mineral' has a rather precise meaning. It is a naturally occurring substance with a specific chemical composition. According to this definition, very few metals can accurately be called minerals. Most occur in nature in combination with other elements. The copper we use, for example, originates in sulphide or oxide minerals. It is the business of mineral processing to extract the wanted element, copper in this case, from these minerals. Economists use the term 'mineral' much more loosely to cover pretty much any inorganic (non-living) substance that comes out of the ground. In its broadest definition, it includes oil, although confusingly it is sometimes used to signify minerals other than oil – as in 'oil and minerals'. Strictly speaking, one should always qualify the term to make clear exactly what it is one is talking about, as in 'non-fuel minerals' (to exclude hydrocarbons, oil, gas and coal) or 'hard minerals' (to exclude oil and gas but to include coal), but this is cumbersome. For present purposes, the term 'mineral' is used to refer to hard minerals, whether processed or unprocessed.

As for the industry which mines these things, it is useful to provide some broad idea as to how the industry is structured and what are the major groupings within it. Table I.1 is based on a classification compiled by the International Council on Mining and Metals (ICMM) in 2012. The cut-off points for the various categories are necessarily somewhat arbitrary. The first group, the mining 'majors', is a rather wide group which includes the important subset of very large companies commonly referred to as the global diversified miners (GDMs), along with a number of the larger single country or single-commodity producers, including state companies. At the other end of the table, artisanal miners are a bit different from the others listed since many of them operate within the informal sector, which is to say they are largely or wholly unregulated. The World Bank estimates that there may be 15–20 million artisanal miners operating across 30 countries.[3] Although they play a significant role in the supply of certain minerals such as tantalum and tin and provide livelihoods to a large number of people, their activities are for the most part beyond the scope of this book. As of 2013, the total market capitalisation of the world's largest one hundred listed mining companies was approaching \$1200 billion, this representing just under 2 per cent of global stock market value.[4]

With regard to country classifications, the book generally follows the International Monetary Fund (IMF) convention of distinguishing between 'advanced economies' and 'emerging market and developing economies'. However, to provide for consistency over time, the 'advanced' grouping is taken to include only the 'old' Organisation for Economic Co-operation and Development (OECD) countries, namely North America (Canada and the United States), Western Europe, Japan, Australia and New Zealand. For brevity, the 'emerging market and developing economies' are often referred to simply as 'emerging market economies' or 'emerging economies'.

Table I.1 Structure of the mining industry, 2012

Company category	Approximate asset base	Number of companies	Comment
Majors	>$3 billion	150	Global and large single-country companies
Intermediates	$1–3 billion	450	Smaller multi-mine companies
Juniors (producers)	$0.5–$1 billion	1500	Companies with one mine
Juniors (exploration)	$5–$500 million	2500	Exploring but not producing
Juniors (starts up)	<$5 million	1500	Seeking exploration funding
Artisanal miners	Small	Millions	Operating in informal sector

Source: ICMM (2012).

Values throughout the book are presented in terms of US dollars ($) and cents (c), unless otherwise specified. 'Bn' signifies a billion, or a 1000 million. GDP data are generally based on purchasing price parities (PPPs). While PPPs are not without their problems, commodity use tends to correlate better with PPP-based GDP than with that based on prevailing exchange rates. Weights of industrial minerals are expressed in tonnes (t), also known as metric tons, with 'kt' signifying a thousand tonnes and 'Mt' a million tonnes. In quoting metal prices, occasional use is made of the pound (lb). Weights of precious metals are expressed in tonnes or in troy ounces (oz). Sources for data used in tables and figures are given in abbreviated form under the relevant table or figure. Full references can be found in the 'Data Sources' section at the end of the book.

The approach of the book is essentially narrative. Inevitably, many issues have been skated over superficially and some complex issues subjected to crude generalisation. Although a book on economics, it is written from the perspective of an institutional economist and goes lightly on theory. For those wanting to delve more deeply into the mechanisms of commodity markets and mining industry economics, there are better places to look.[5]

Back in 2005, a year or so after I joined Norilsk Nickel, one of my younger colleagues remarked to me how lucky we were to be in mining, given how easy an industry it was in which to make money. The remark made me pause for thought since it ran counter to everything I had learned about the industry and how it had worked over the preceding quarter century. That, probably more than any other, was the moment I realised something fundamental had changed and that the job at hand was no longer to try to shoehorn what was going on into our existing model of how things worked but to figure out the characteristics of the new world into which we were moving and what they meant for the future.

1
The Barren Years

Are they still doing that?

It was the autumn of 1999, and Hugh Morgan, the forthright head of Western Mining Corporation, was entering the United Kingdom to deliver an address at the London Metal Exchange (LME) dinner, one of the biggest events in the global corporate calendar. In response to an immigration officer's enquiry regarding his occupation, he replied that he worked in the mining industry. The officer eyed him quizzically and asked, 'Are they still doing that?'[1]

The theme of Hugh Morgan's address that evening was the contribution that technology could make to improve productivity in mining. However, this was not the sort of technology the world was then interested in. The sort of technology that was grabbing the public's imagination was ICT, information and communications technology, more specifically the fast-expanding Internet and World Wide Web. In what became known as the 'dotcom' bubble, the index of the NASDAQ, the US stock exchange specialising in technology stocks, soared to a peak of 5049 in early March 2000, more than double its level a year before. Meanwhile, mining stocks languished, unwanted and unloved. The industry was deeply unprofitable, and its future looked bleak. Between 1995 and 2000, mining's share of global stock market values declined from 5 per cent to 1 per cent. In 1999, the entire quoted global mining industry had a market capitalisation of around $300 billion. In December of that year, the value of one technology company, Microsoft, was double that, $620 billion. This did not bode well for the industry entering the new millennium. The public commonly perceived the industry as dark, dirty and dangerous, certainly no place for the young and ambitious.

Disappearing demand

Several factors contributed to the dire condition of the mining industry at this time.

The first factor was the trajectory of global economic growth. The demand for mineral raw materials is highly sensitive to the rate of economic growth and to changes in the rate of growth. The global economy had rattled along fitfully through the 1950s to the 1970s, supported by a vibrant US economy and the post-war reconstruction of Europe and Japan. However, from the 1970s, things slowed up radically. Over the period 1950–1975, global GDP (Gross Domestic Product) had achieved an average growth rate of 4.8 per cent a year. Through 1975–2000, it managed just 3.0 per cent.[2]

There were two key reasons for the slowing of global growth. The first reason was the deep economic recession that followed the oil crisis of the late 1970s. This was in part caused by the tight monetary policies applied at the start of the 1980s to squeeze inflation out of Western economies. The second reason was the economic slowdown and eventual collapse of the Soviet Union. Between 1990 and 2000, the economies of the former Soviet Union (FSU) shrank in size by over a third.[3] Both these developments had a severe negative impact on global demand for minerals.

There was, however, a bit more to it than this. Along with the slowing in the rate of economic growth, changes were taking place in the *nature* of that growth. As Western economies emerged from the period of post-war reconstruction, their growth was becoming less material intensive. This partly reflected the fact that the focus of growth was shifting away from investment towards consumption. Investment as a percentage of GDP (expenditure based) in Japan had peaked at 36 per cent in 1973. By 2000, the share had fallen to 26 per cent. At the same time, the focus of consumption in Western industrial economies was shifting from the production of things to the provision of services. In 1980, the share of services in Japan's GDP (output based) was 58 per cent; by 2012, it had risen to 73 per cent. The equivalent figures for the United States were 64 per cent and 79 per cent.[4] The share of manufacturing in these economies displayed a corresponding decline. Growth, in short, was moving to activities which simply used less stuff.

In addition to these changes in the composition of growth, products were themselves becoming less material intensive. Copper tube used for plumbing was being produced with narrower gauges and thinner walls, thereby reducing the amount of copper required to produce it.

Miniaturisation was cutting into the market for copper wire. Improvements in electro-galvanising meant that galvanised steel could be produced with thinner coatings of zinc. Lead was being phased out of use in petrol as an anti-knock agent. In response to the oil shocks of the 1970s, drivers in the 1980s switched to buying smaller cars, reducing the amount of steel they used. The development of higher quality alloy steels was resulting in the light-weighting of architectural structures.

As a consequence of these two trends – the declining product intensity of GDP and the declining material intensity of products[5] – demand for many mineral commodities was being eroded. 'Intensity of use', a concept used by economists to measure the amount of a commodity employed for each unit of GDP, appeared to be in long-term decline. Between 1970 and 2000, the intensity of use of nickel declined by 25 per cent and that of steel by 47 per cent (see Figure 1.1).

According to some observers, this 'dematerialisation' of economic activity was the pattern for the future. The smokestack industries of the old economy were giving way to a digital 'new economy'. The central idea of the new economy was that the generation and exploitation of knowledge were going to play an ever-increasing part in the creation of

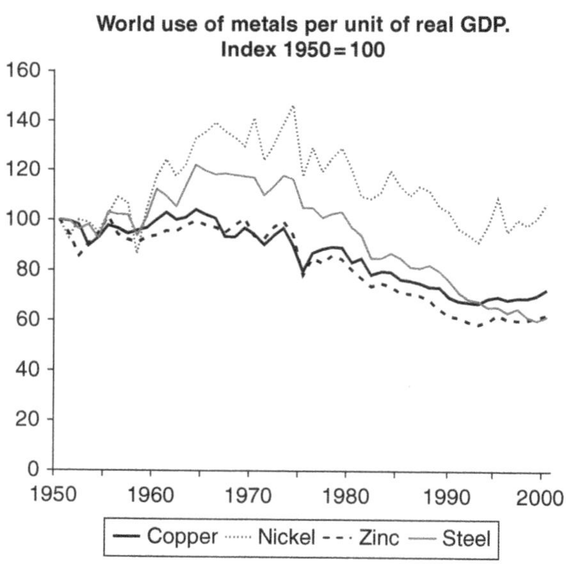

Figure 1.1 Intensity of use of selected metals, 1950–2000
Source: Maddison (2006), IMF, Metallgesellschaft, WSA.

wealth. The fundamental natural resources in this world were not things dug out of the ground but human creativity and intelligence. Wealth would increasingly be embodied in intellectual property rather than in things you could drop on your foot. The vehicles for this knowledge were microchips, computer software, digital data transmission, multimedia and mobile telephony. Moreover, it had a particular property – it was increasingly 'weightless'.

One of the exponents of the 'weightless economy' explained it thus:

> During the past half century or so, technological and economic changes have meant that we produce things of much less bulk: transistors rather than vacuum tubes, fibre-optic cables or satellite broadcasting rather than copper wire, plastics rather than metals. Materials have changed and miniaturisation has been pervasive.... The value in our economy – whatever it is we are prepared to pay money for – has less and less physical mass. Whether it is software code, genetic codes, the creative content of a film or piece of music, the design of a new pair of sunglasses or the vigilance of a security guard or helpfulness of a shop assistant, value no longer lies in three-dimensional objects in space. We will pay for amusement, for style, for convenience, for speed, for creativity, for beauty – but when it comes to things, commodities, we have turned into skinflints, and want the cheapest possible.[6]

While there was clearly something in this argument, it was a very partial perspective. It overlooked, for example, the investment in basic materials which the digital economy required for the fabrication of computers, switchgear, routers and servers, and the power stations, fuels and transmission networks required to keep them running. Power brownouts in California, the heart of the new economy, during 2000 and 2001, revealed that some of the new technology plants were consuming the power of small cities. Also, it ignored the structural fabric of the weightless economy. The new economy did not exist in the ether but was housed in offices, trading floors, shopping malls, multiplexes and homes, all of which required substantial amounts of materials for their construction. Those made rich by the dotcom boom were ploughing their money back into bigger, power-hungry homes and newly built health clubs and resorts.[7] Between 1985 and 1998, the copper intensity of the average US home increased from 0.63 kilograms per square metre to 1.02 kilograms per square metre according to the Copper Development Association.

More importantly, it was very much an advanced economy perspective. It took little account of the billions of people around the world who did not have the material possessions and facilities enjoyed by those living in the advanced economies. Many of these people were very interested indeed in material intensive products such as houses, refrigerators and cars. At the time, two-thirds of the world's population did not have access to a telephone and one-third did not have electricity. It was true that Western economies had in the past dominated the global economy and, along with this, the consumption of mineral raw materials. Even in the late 1990s, when the above-mentioned quotation was being penned, they still accounted for around two-thirds of the global consumption of metals. But this was beginning to change – and change dramatically.

Too much of everything

The process of dematerialisation identified by the exponents of the weightless economy posed a major dilemma for companies in the mining industry, an industry whose future depended on continuing growth in demand for material products.

The industry was slow to grasp the significance of the oil shocks and of the economic recession that followed in 1980–1982. The general presumption was that this was a hiccup, and that once it had passed, the rates of growth which had preceded the crisis would resume. Accordingly, companies retained capacity which was surplus to requirement rather than close it down permanently and remained open to the idea of investment in mines which might be needed as and when the recovery of demand finally arrived.

This belief that demand growth would, in time, pick up to something close to pre-oil crisis rates found support in academic and government research. One such piece of research, published in 1983, was carried out by the Institute for Economic Analysis (IEA) at New York University under the direction of Nobel Laureate, Wassily Leontief. Using the input–output technique which Leontief had pioneered, the study made a series of projections for the demand for mineral raw materials to the year 2030.[8] For the period 1970–2000, it projected annual average growth in world demand for copper, nickel and zinc, at 4.4 per cent, 2.7 per cent and 3.5 per cent, respectively. This compared with actual rates of growth for copper, nickel and zinc demand over the period 1950–1980 of 3.9 per cent, 5.1 per cent and 3.7 per cent. In other words, the IEA was projecting a rate of growth in demand for copper which was slightly above the recent historical experience, a rate for nickel which was somewhat below and a rate for zinc which was pretty much in line. Nickel aside, the impression conveyed by these projections was one of continuity. In similar vein, the US government agency, the US Bureau of Mines, was at this time projecting annual growth in demand over the period 1970–2000 at 3.9 per cent for copper, 4.3 per cent for nickel and 2.1 per cent for zinc.[9]

Against this backdrop, it was not so surprising that, as the global economy started to recover in 1983, the industry determined that normal service had once more resumed and opened the taps on production. It was too much, too soon. Although global copper demand turned positive in 1983 and 1984, the market was overwhelmed by new supply. Stocks of copper on the metal exchanges of New York and London, which had collectively stood at 300,000 tonnes in early 1982, soared to over 800,000 tonnes by the end of 1983. Prices over the same period slid from $1600/tonne (73c/lb) to around $1400/tonne (64c/lb). Although greater restraint by copper producers meant that stocks declined thereafter, and kept on declining, it wasn't until the second half of 1987 – nearly four long years later – that the copper price finally began to stage any sort of recovery.

The condition of the iron ore market at this time was, if anything, even worse. The vast iron ore deposits of the Pilbara in Western Australia had been opened up during the 1960s to meet rising demand from Japan's fast-growing steel industry. Between the early 1950s and 1973, Japanese steel production grew at over 15 per cent a year. Lacking any iron ore resources of its own, this gave rise to a burgeoning demand for seaborne iron ore. However, with the oil shocks in the 1970s, Japan's growth in steel production came to an abrupt halt, and so too did its requirement for increased tonnages of imported iron ore. Having reached a peak of 119 million tonnes in 1973, Japan's annual steel output slipped back into the range 100–110 million tonnes, where it remained for the next 20 years. Mine projects which had been committed on the assumption that the earlier higher rates of growth would persist, found they had nowhere to place their product.

Adding to the problems of oversupply in the industry, Brazil was in the process of making a substantial addition to world iron ore supply. Vast deposits of high-quality iron ore had been discovered in the Amazon Basin in the 1960s, and plans were developed by the state company Companhia Vale do Rio Doce (CVRD) to build what would become the world's largest iron ore mine at Carajás. In 1978, work started on the 890 kilometre railway that would bring the iron ore out of Carajás to the coast at Ponta da Madeira from where it could be shipped. By 1985, work on the mine and the associated infrastructure was complete and production at the mine commenced.

The timing of Carajás could scarcely have been worse from the perspective of the global market balance. Although the target of Brazil's iron ore sales at this time was Europe rather than Japan and the Asian market, the European steel industry was in no better condition to absorb large additional quantities of iron ore than was Japan. Having peaked at 156 million tonnes in 1972, steel production in the European Community had slumped to 128 million tonnes by 1980 and the industry was operating with a capacity utilisation rate of only 60 per cent.[10]

The world was simply awash with iron ore, and prices reflected this. The seaborne iron ore market had its own distinctive way of establishing prices. Typically, the Australian iron ore producers negotiated with leading Japanese steel mills while Brazilian and Canadian producers negotiated with leading European steel mills. By convention, the price of the first settlement was adopted as the 'benchmark' price by buyers and sellers in both Pacific and Atlantic markets for the following year's shipments.

Between 1982 and 1988, the benchmark price of seaborne iron ore fell every year, ending 32 per cent below its starting point. Observing his port disappearing under mountains of unwanted iron ore stocks, one senior Australian manager lamented, 'You can't eat iron ore'. During the negotiations on 1987 prices, the Japanese steel mills alleged that one of the main Australian producers, Hamersley Iron, was trying to stem the fall in prices by orchestrating a coordinated producer response, in contravention of competition law. Accordingly, the Japanese mills that year rewarded Hamersley's principal competitor, BHP, with additional contract tonnages, albeit at a 5 per cent reduction in price. The irony of the fact that the steel mills in Europe and Japan openly coordinated their stances in contract negotiations was apparently lost on the buyers.

The North American iron ore market was in no better condition than those in Asia and Europe. Having substantial resources of iron ore in the northern states (Minnesota and Michigan) and across the border in Canada, US steel producers had no need for supplies from the seaborne market. However, declining steel demand was diminishing the need for local iron ore among steel makers around the Great Lakes. In addition, the position of the large integrated steel mills using blast furnaces and iron ore was being challenged by the development of small producers using electric arc furnaces (EAFs) and steel scrap as feed. The availability of scrap and power being the most important inputs for these producers, they were able to locate themselves away from the traditional steel-producing areas. Not requiring the infrastructure and scale of integrated steel plants, and being free of the constraints of using unionised labour and the crushing pension obligations of the major steel producers such as US Steel and Bethlehem Steel, these EAF producers were inherently lower cost than their larger brethren. In consequence, North American producers of iron ore suffered badly, the largest of them, Cleveland Cliffs, being pushed to the brink of bankruptcy and into a debt restructuring programme in 1987.

Although mineral commodity markets rallied towards the end of the 1980s, the pattern had nonetheless been set for a number of years to come, and by the early 1990s, the industry was once more grappling with conditions of weak prices and chronic oversupply. Miners continued to produce and invest for a higher level of demand than was required. Any upturns in the industry's prospects were quickly overwhelmed by new supply. Although industry leaders frequently remarked that this situation could not go on and that sooner or later prices would have to rise to levels which provided some sort of return on capital if companies were to have an incentive to invest, by using such 'incentive' prices for the purposes of evaluating their projects they were helping to ensure that the higher prices never arrived. The tendency to use unrealistically high prices in their project assessments can probably be attributed in part to the stubborn optimism of the mining industry, while the tendency to overproduce may have owed something, at least for some commodities such as iron ore and nickel, to a destructive pursuit of market share among producers. It also seems likely that a part was played by 'social production', that is to say, production from operations, often under state ownership, which was kept operating in order to preserve employment or flows of foreign currency. Whatever the reasons, it was a condition one CEO graphically described as 'supply incontinence'.

A final challenge for the mining industry during these years arose from the collapse of the Soviet Union. As one of the two great global powers during the Cold War, the Soviet Union was a major producer and consumer of mineral commodities. In 1970, the Soviet Union was the world's largest producer of iron ore and platinum group metals and the second largest producer of aluminium, copper and gold. Although its growth slowed during the 1980s as more and more economic resources were directed towards maintaining its military capability, the Soviet Union was still a very large mineral producer at the time of its disintegration in 1991. In the economic collapse that followed, numerous factories were closed and military spending was dramatically curtailed. Accordingly, enormous amounts of mineral raw materials, which would otherwise have gone into domestic manufacturing, became available for export. In addition, a large amount of obsolete plant at defunct factories was dismantled and exported as scrap to generate foreign currency.

Thus, at a time when Western markets were having difficulty absorbing supplies from their own sources, more and more mineral products were pouring out from the states of FSU. The surge in exports took markets by surprise, not only because of the speed at which it occurred but also because of its scale. In fact, it soon became apparent that Western analysts had been underestimating the scale of Soviet production of minerals and metals during the Cold War. While the most authoritative estimates had Soviet production of aluminium in 1990 in the region of 2.5 million tonnes, it turned out to have been near 3.5 million tonnes, one million tonnes higher. Similarly, while Western analysts generally had Soviet production of nickel running at around 200,000 tonnes a year, it transpired that production was actually around 260,000 tonnes a year.

In any event, exports of aluminium soared from a few hundred thousand tonnes a year in the late 1980s to over 2.5 million tonnes by the late 1990s with predictably negative consequences for prices and for Western producers of the metal. Indeed, so severe did things become that the governments of aluminium-producing countries (including Russia) agreed a memorandum of understanding (MoU) in 1994, permitting producers to cut back their production on a coordinated basis without invoking censure by competition authorities in the United States and Europe. Exports of refined nickel from Russia more than doubled over the same period to over 200,000 tonnes a year. Exports of copper scrap increased from virtually nothing to half a million tonnes a year in 1997 and 1998 before the Russian authorities imposed export controls so as to prevent their own smelters, which wanted to be able to export value-added copper products, being deprived of raw materials.

As a consequence of these persistent conditions of oversupply, mineral commodity prices through the two decades of the 1980s and 1990s remained under almost constant pressure. Economic theory contends that, in competitive markets, commodity prices will reflect the costs of the marginal producer, that is, the last (highest cost) producer required in production to balance the market. In the long run, these costs will include fixed and variable costs, in effect both capital costs and operating costs. In the short run, however, the only costs relevant to the determination of prices are variable, or operating, costs. These operating costs, theory suggests, will provide a floor below which prices should not fall. The experience of these years suggested something different. As prices pressed down on producers' costs, these costs kept falling. They were not serving as a floor. In addition, it was becoming apparent that prices falling below operating costs provided no guarantee that producers would shut down and reduce the available supply. Because of the capital intensive nature of the industry and the high social costs of closure, the 'avoidable' costs of producers were turning out to be below operating costs, leading some producers to stay on in production even while losing money. The outcome of all this was a spiralling down of costs and prices. As prices pressed down on costs, cost reductions were making possible further reductions in prices.

The impact of this process on mineral prices was severe. Aside from a short-lived run-up in prices in the late 1980s, they generally moved sideways throughout the two decades (see Figure 1.2). The price of copper in 1978 averaged $1363/tonne (62c/lb); in the first half of 1999, it averaged slightly more, $1436/tonne (65c/lb). In 1979, the price of nickel averaged $6000/tonne ($2.72/lb); 20 years on, in 1999, it was almost exactly the same. What went for non-ferrous metals went also for iron ore. The nominal price of traded iron ore in 2000 was almost exactly the same as it had been in 1980. And yet, during this period, dollar prices

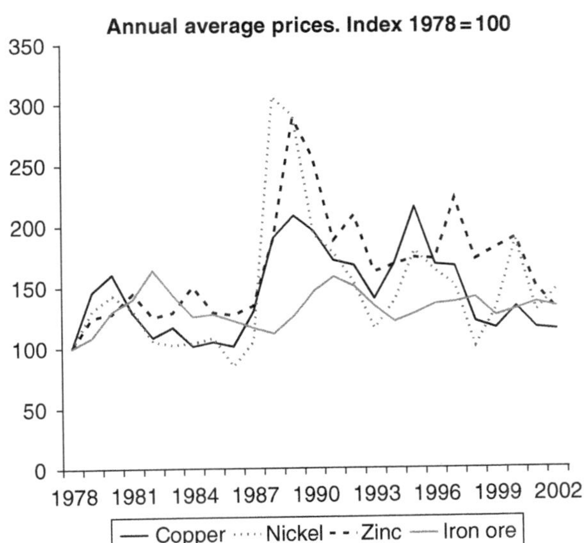

Figure 1.2 Metal prices, 1978–2002
Source: LME. The iron ore price is the Hamersley Iron fob export price for fines.

of goods generally, as measured by the US GDP deflator, had almost doubled, implying that the price of these mineral commodities in real, inflation-adjusted terms had effectively halved. Across the mineral spectrum, prices in real terms appeared to be on a declining trend. This was a deeply challenging environment for any industry to operate in and a profoundly discouraging one for anyone thinking about investing in the sector.

An industry in distress

A story is told of an exploration manager reporting to the board of a gold mining company at a time when copper prices were at rock bottom in the 1980s: He says he has good news and bad news to report. The board asks for the bad news first.

> 'Unfortunately, we haven't found any gold'.
> 'And the good news?'
> 'We didn't find any copper either'.

The mining industry has a rich history of black humour, but it was never so pertinent – and the need for a sense of humour so acute – as during these years.

As time progressed, it was becoming increasingly apparent that the rates of growth in metal demand which had been projected in the years following the oil price shocks, were not going to be fulfilled. The IEA, as noted above, had forecast annual average growth of 4.4 per cent for copper consumption, 2.7 per cent for nickel and 3.5 per cent for zinc over the years 1970–2000. The reality was that copper and nickel consumption grew during these years at 2 per cent and zinc at 1.4 per cent. The consumption of iron ore, which the IEA had forecast to grow at just over 4 per cent a year over this period, grew scarcely at all.

These were not propitious conditions for the mining industry. The low rates of growth deprived the sector of dynamism, while the persistent tendency for producers to overproduce suppressed margins and profitability. Investors voted with their feet, and share prices slumped. The idea of 'weightless' products and 'new economy' share offerings seemed much more appealing. An analysis conducted in 2000 for the 27-year period 1973–1999 showed that the resources sector as a whole had achieved a reasonably respectable 8 per cent average real rate of return on capital employed (see Figure 1.3). However, this achievement was entirely down to the oil and gas sector, which, in 1999, had a market

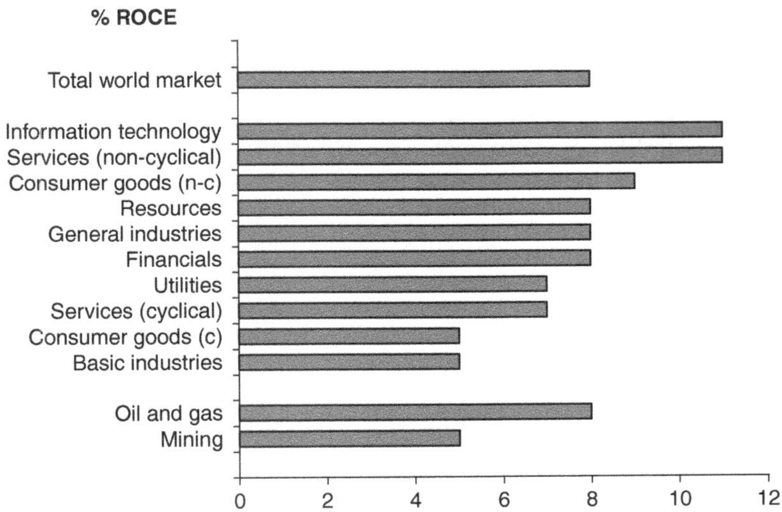

Figure 1.3 Profitability by industry sector, 1973–1999
Source: Humphreys (2001).

capitalisation some eight times that of mining. Real returns for mining were a much more modest 5 per cent, positioning it alongside so-called basic industries, such as steel, construction, forestry and chemicals, as the worst performing industry sector.

Not surprisingly, young people could see less and less a future for themselves in such an economically unappealing industry and were abandoning it in favour of other more dynamic sectors. Mining schools had capacity designed for another era, and many were closing down. The industry was deeply demoralised and floundering, a sad commentary on a sector that had historically been famed for its unrelenting optimism.

To survive in this hostile environment, the industry had to adapt, which it did through cost reduction and restructuring.

Mining is, in economic terms, a relatively simple industry. Because prices are largely outside the control of producers and because there is virtually no scope to achieve premiums from branding or clever marketing, the profitability of producers is almost entirely a function of their costs. Accordingly, this is the first area that producers are obliged to address when the going gets tough.

Certainly, producers became very creative with respect to costs during these years. Labour costs were reduced throughout the industry by lay-offs and improved working practices. In some regions, notably the United States and Australia, de-unionisation facilitated workforce reductions and increased worker flexibility. Operating costs were reduced through the systematic exploitation of economies of scale and the use of larger and larger equipment. The largest available truck in the late 1970s had a carrying capacity of around 150 tonnes. This had doubled to over 300 tonnes 20 years later. The deployment of new technologies, such as the application of solvent extraction-electro winning (SXEW) for the production of copper, also played a part. Weak energy prices permitted companies to reduce their energy costs. In the aluminium industry, smelters were able to negotiate risk-sharing agreements with their power suppliers, which meant that their power costs declined along with the aluminium price, thereby protecting their margins. The visible stress of the industry enabled producers to bring pressure to bear on their suppliers of equipment and parts to achieve lower purchase prices.

Unfortunately, because lower costs were simply leading prices lower, these policies were not in themselves sufficient to increase industry profitability. The industry also had to rein in its investment and to restructure.

The part of the industry worst hit by this was the US mining sector. As a result of its economic dominance in the post–World War II era, as well as its large size and rich mineral endowment, the United States boasted many of the largest names in the mining industry; Amax, Anaconda, Asarco, Cleveland Cliffs, Cyprus, Doe Run, Kennecott, Magma, Newmont, Peabody and Phelps Dodge. With oil prices riding high in the late 1970s, some of these companies had been acquired by US oil companies looking to diversify. Anaconda was acquired by Arco, Cyprus by Amoco, Amax by Chevron and Kennecott by Sohio. However, nothing could protect these companies from the gales that were sweeping through the industry in the 1980s. With ageing assets, bloated workforces, insufficient geographic diversification and a strong US dollar (which was inflating US production costs relative to those elsewhere in the world), the industry was in deep trouble.

As a result of mine closures, massive lay-offs, the use of new technologies such as SXEW and a relentless focus on cost reductions, the industry fought back. The average cost of producing copper in the United States was reduced by 28 per cent between 1982 and 1987, from \$1740 a tonne (79c/lb) to \$1255 a tonne (57c/lb).[11] Notwithstanding this achievement, the industry had lost much of its interest for US investors as well as its status as the leader of the global mining industry. In a famous cover, in December 1984, the US magazine *Business Week* proclaimed 'The Death of Mining'. Two years later, in 1986, a long-serving senior executive of Asarco, Simon Strauss, published *Trouble in the Third Kingdom*, chronicling the challenges then facing the mining sector. The book starts with a discussion of the perceptions of scarcity, which had shaped the public discourse on mining in the immediate post–World War II years, and the heroics of the mining industry in rising to and meeting the challenges posed by strongly rising mineral demand during these years. The author then records, with a palpable sense of regret – and not a little bafflement – the manner in which, in the 1980s, concerns of scarcity had given way to those of over-abundance and how the status of the mining industry had been diminished. In Strauss's words:

> [T]he post-war years brought prosperity to the minerals industry – The Third Kingdom. Yet now that Kingdom is in trouble.... after 30 years of rapidly expanding consumption, the rate of growth has perceptibly slowed for most minerals.... Substantial over-capacity exists in many mineral commodities. In part, this is the consequence of new projects started prior to 1974 when markets were expanding but completed since then in a time of market stagnation.... Periods of sustained over-production have led to accumulations of excessive inventories. As a consequence, prices have been weak.... Heavy financial losses were sustained by mineral producers during the severe recessions of 1975 and 1981–82. Numerous mines and processing plants closed, workers lost their jobs, and the economies of entire regions have been undermined.[12]

These conditions, and the need to rationalise the remaining productive capacity, led to the wholesale restructuring of the US mining industry and in several instances the loss of control of these once-dominant companies to foreign owners. In 1989, UK-based RTZ (now Rio Tinto) acquired Kennecott as part of a package of assets owned by oil giant BP. The British-American conglomerate Hanson bought US coal producer Peabody in 1990. In 1993, Amax and Cyprus Minerals merged to form Cyprus Amax; then, in 1999, the resulting company was absorbed into fellow US miner Phelps Dodge. In 1993, Asarco was acquired by Mexican copper producer Grupo Mexico, and in 1996 the Magma Copper Company fell to Australia's largest mining company, BHP. By the end of the decade, the US industry was scarcely recognisable from what it had been 25 years previously.

Producers elsewhere in the world were not finding things much easier. London, for long a major centre of mining finance, by the end of the

1980s featured only one miner (RTZ) in the FTSE100, the London Stock Exchange's largest 100 companies by market capitalisation. In continental Europe, the depletion of local mines and the loss of supplies from former colonies had left a substantial metallurgical industry with insufficient raw materials. Most steel producers were wholly dependent on imports for their supplies of iron ore, while companies processing non-ferrous metal ores, such as Penarroya and Société Le Nickel of France, Metallgesellschaft and Preussag of Germany, Union Minière of Belgium and Outukumpu of Finland, were also largely dependent on imported raw materials. Because of the amount of capital tied up in them, large metallurgical plants require high levels of capacity utilisation to cover their costs, and mine closures around the world were making this a challenge. Attempts by some of the companies to invest in mining ventures overseas to secure their own captive supplies were largely unsuccessful, and many European metal producers were forced into mergers with other European producers and with foreign raw material suppliers in order to survive.

Japan was in an even worse position than Europe with respect to its lack of domestic mine supplies. It had virtually none. In an attempt to gain direct control over supplies of raw materials, most of the big conglomerates, such as Mitsui, Mitsubishi and Sumitomo, had set up mining companies with a view to them investing in mines overseas to supply their domestic smelters and refineries. In this project, they were actively supported by the Ministry of International Trade and Industry (MITI – now part of the Ministry of Economy, Trade and Industry (METI)) and agencies such as the Metal Mining Agency of Japan (MMAJ – now part of Japan Oil, Gas and Metals National Corporation (JOGMEC)). However, these companies found it hard to compete on an equal footing with established international miners, and Japan's policy adapted over time away from finding and developing mines towards that of investing in other companies' mining projects in return for securing off-take agreements.[13]

A glittering exception

An important exception to the tale of gloom that enveloped mining in these years was gold. The US administration severed the link between the dollar and gold in 1971 (it had previously been set at $35 an ounce), after which the price of gold moved steadily upwards. Receiving a further boost from the loose money policies and high rates of inflation in the advanced economies in the late 1970s, gold soared to unprecedented levels. After recording an average price of $110/oz in August 1976, gold edged up to $206/oz in August 1978 before exploding to an average of $673/oz in January 1980. Although prices came off these rarefied heights during the course of 1981 and 1982, they remained at what were historically high levels for the next 15 years, creating a unique opportunity for gold miners.

As of 1975, South Africa was by far and away the world's largest producer of gold, accounting for almost 60 per cent of all gold mined worldwide. However, South Africa was not well placed to respond to the upsurge in the price of gold. Its mines, already deep and difficult to work, were getting progressively deeper and costlier. Moreover, the prevailing system of apartheid in South Africa meant that many international investors were reluctant to make large investments in the country. Accordingly, other parts of the world were better positioned to take advantage of the situation. As gold prices surged, gold deposits which had previously been considered economically marginal in more investor-friendly regions were being priced into profitability. Foremost among these countries were the United States, Canada and Australia. In the United States, the so-called Carlin Trend in Nevada saw a huge inflow of investment while something similar was occurring in the Kalgoorlie area of Western Australia. In 1991, it is estimated that these three countries accounted for around 60 per cent of global mineral exploration activity, with the majority of this activity being directed towards gold.[14] South Africa's share of world gold production meanwhile slumped to 26 per cent in 1990.

For an industry which had little else to celebrate, developments in gold were welcome, but even these were not to last. By the latter part of the 1990s, gold prices were beginning to weaken and some of the big gold producers were systematically selling their gold production forward. Canada's Barrick Gold grew to become the largest and most profitable gold producer in the world on the back of this strategy. Other companies adopting the strategy included Newmont Mining, AngloGold Ashanti, Goldcorp and Gold Fields. While the forward selling of gold may have made sense for an individual company in a falling price environment, it nonetheless represented a de facto vote of no confidence by gold producers in the price prospects for their product and also served to mobilise otherwise dormant holdings of gold, adding further downward price pressure. Banks taking the other side of a miner's forward sale typically borrowed gold bullion from central banks for a small rental fee. They then sold the gold into the market and put the resulting funds on deposit so as to be able to buy back the gold and restore it to the central bank when the position matured. In the process, they were adding to the supply of gold in the market.

Worse than this, many central banks, particularly those in Europe, had determined that in the low-inflation environment then prevailing, holding large amounts of gold as part of their reserves was no longer as important as once it had been and began to sell off their gold. Such official sales of gold were 130 tonnes in 1994 but rose steadily to 477 tonnes by 1999 before peaking at 674 tonnes in 2005. The UK government famously picked the bottom of the market to sell half its gold reserves, a total of 375 tonnes, in a series of auctions between 1999 and 2002. Not much helped by the advance signalling of these sales, the average price achieved for the gold was a very modest $275/oz.

As the gold price began to crack, the love affair of the mining investment community with gold inevitably began to fade. Since gold had for years been the only interesting mining play around, it had become the target for some unscrupulous operators, and a number of mining investors had had their fingers burnt by companies engaged in fraudulent activity. The apogee of the gold stocks' boom was reached in early 1997 before the market was rudely brought down to earth by the implosion of a Canadian gold exploration company called Bre-X. This small Calgary-based company had reported a sensational gold find in the rainforests of Borneo in Indonesia in 1995. The scale and richness of the find grew continuously and spectacularly through 1996. However, in March 1997, shortly after the chief geologist working on the project mysteriously fell out of a helicopter near the deposit, US miner Freeport-McMoRan, which had been assessing the deposit with a view to buying it, revealed that it contained 'insignificant amounts of gold'.[15] The effect on the Bre-X share price was swift and brutal, and in the aftermath the prices of many other junior mining stocks plummeted. Serious damage had been done to investors' confidence in the reporting standards of listed companies and to public perceptions of the integrity of the mining industry.

By the end of the 1990s, when Hugh Morgan was addressing the LME dinner, the industry was in a bad place. Prices of non-ferrous metals were in nominal terms pretty much where they had been 20 years before and in real terms a whole lot lower. The boom in gold, the only bright star in the minerals firmament, had finally turned to bust. Battered by years of poor profitability, talk in board rooms was all about cost reduction and capital efficiency – 'making assets sweat' in the parlance of the day. There was a growing focus on how new technologies could be used to drive down procurement costs, for example, through the use of Internet-based Dutch auctions. The spirit was one of fatalism. The industry was settling down for a long grey slog in a low-growth industry offering only modest returns.

Not surprisingly, given these conditions, there was little appetite for investment in new mining projects. Lack of confidence in the outlook for the sector coupled with low share prices had made companies risk averse. Although many emerging countries had opened up their economies to inward mining investment during the late 1980s and 1990s – over 100 countries rewrote their mining laws during these years with a view to attracting foreign investment[16] – history shows that, in fact, most investment outside the advanced-economy countries during these years went to a very small and select group of emerging-economy countries, namely Chile, Peru and Indonesia.[17] Russia had notionally opened itself up to inward investment in mining, which it certainly needed after many years of neglect and underinvestment, but the risks of going into the country were considered too high for most Western companies. Many took a look, but few stayed. As if it was not challenging enough already, in 1998 Russia defaulted on its foreign debt obligations and suffered a major financial crisis.

To be able to operate more effectively under these demanding conditions and to rationalise the industry's excessive productive capacity, some of largest companies in the industry sought opportunities for consolidation. In effect, they were circling the wagons. The extensive rationalisation of the US mining industry has already been noted. But the same was also going on elsewhere. UK-based RTZ merged with its 49 per cent Australian subsidiary CRA in 1995, subsequently naming the joint entity Rio Tinto. In 2000, the company took over North, an Australian iron ore producer. South Africa's largest company, Anglo American, moved its primary listing from Johannesburg to London in 1999 and absorbed back into its main structure the assets of Minorco, a company it had set up to look after the company's non-South African interests during the apartheid era. In 2000, former Shell mining subsidiary, Billiton, acquired Canada's Rio Algom before itself merging with Australia's BHP in 2001 to form BHP Billiton. In the same year, Vancouver-based Teck acquired fellow Canadian zinc–lead producer Cominco. The aluminium industry also underwent a wholesale rationalisation at this time with the world's largest aluminium producer Alcoa taking over fellow US producer Reynolds in 2000 and Canada's Alcan acquiring Swiss-based Algroup in 2000 and French producer Pechiney in 2003.

A crisis of trust

On the night of 19 August 1995, following heavy rainfall, a breach occurred in a dam at the Omai gold mine in Guyana, resulting in the spillage of large quantities of mine waste into the Omai and Essequibo Rivers. Although the mine's operators, Canadian company Cambior, moved swiftly to stem the leak, over the course of the next 100 hours four million cubic metres of cyanide-containing waste were released into the river system. Following protests from residents, who used the river as their primary source of water for drinking and agriculture, Guyana's president declared 80 kilometres of the Essequibo River an environmental disaster zone.

Subsequent investigations revealed that initial claims about the scale of the damage had been vastly overstated. Although a small number of dead fish were found in the Omai River, none were found in the Essequibo River into which the Omai flows. Five days after the accident, cyanide in the Essequibo River had returned to normal, safe levels. Although cyanide can be lethal to humans, as every reader of crime novels knows, it quickly breaks down and becomes harmless when exposed to sunlight. A number of international organisations, including the United Nations, the Organization of American States, the World Health Organization and the Pan American Health Organization were called in to monitor and help remediate the spill's effects. A $2 billion class action brought on behalf of 23,000 local residents against the company was dismissed by Guyana's Supreme Court in late 2006.[18] However, the damage to the industry had been done, and Omai mine's spillage could be added to the long list of high-profile incidents which over the years had served to undermine public trust in the sector.

As the mining industry entered the twenty-first century, it was not just the poor condition of the market which was pressing down on it and giving rise to a siege mentality in its ranks. Concerns about the sector stretched across many other aspects of its behaviour and performance. During the more than two decades in which the mining industry had been struggling with the issues of weak demand and prices, the world had changed dramatically. The Berlin Wall had come down and the Cold War had ended. Countries which had previously enjoyed the patronage of the Soviet Union were being required to find their own way in the world. China was edging out onto the world stage. Information and communications technology had improved dramatically, facilitating access to data and opinion and generally shrinking the globe. Combined with an increased awareness of what was going on elsewhere in the world was a growing concern about how businesses comported themselves with respect to the environment and to the communities in which they operated. For an industry which involves digging large holes in the ground and which has a large presence in emerging economy countries, this was a major challenge.

The nature of mining means that its environmental impacts have always been near the top of the list of public concerns about the sector. Growing environmental activism and improvements in communications had greatly increased the ability of non-governmental organisations (NGOs) to monitor and react to environmental events around the globe and draw the spotlight of public opinion onto mining industry transgressions. In the more economically advanced parts of the world, there was growing resistance to mining because of its large-scale disturbance to land, its visual impacts and the problems associated with its solid and liquid wastes. With growing affluence and increasing pressure from alternative land uses, opposition to mining in the United States and Western Europe was mounting. And even where mining was tolerated, the environmental standards required by mine operators were ratcheting ever higher.

Tailings dams, the dams behind which mining companies accumulate liquid wastes arising from the process of concentrating ores so as to allow sediments to settle out, represent a particular vulnerability for mining, as the Omai case illustrates. Unfortunately, this was not an isolated case. Australian mining major BHP had experienced a tailings dam failure at its Ok Tedi copper mine in Papua New Guinea (PNG) in 1984. This being an area prone to seismic activity, the decision was made not to rebuild the dam but to dispose of the tailings into the Fly River. (BHP subsequently pulled out of the mine claiming that it should never have become involved in it in the first place.) Canadian company Placer Dome experienced a dam failure in 1996 at the copper–gold Marcopper mine in the Philippines, which it managed, effluent from the dam ending up in the Boac River. A further example was provided by a tailings dam breach at Aurul's Baia Mare gold mine in Romania, in 2000. This illustrated another hazard with such events, which is that they are not necessarily confined to one country. The spillage from Baia Mare flowed into the river Tisza across the border in Hungary before ending up in the Danube.

Another key area of sensitivity for mining was its effect on neighbouring communities. Historically, authorisation for mine development in most countries had been made at a national level and based on assessments of what was in the national economic interest. The costs and benefits to the local community were not always properly taken into account. However, as the interest of government in the mining sector began to fade along with the industry's profitability in the 1980s, and it began gradually to disengage from the affairs of the sector, the challenge to mining companies from the state began to be replaced by a challenge from the communities most directly impacted by their activities.[19]

Quite the most dramatic example of the breakdown of company–community relationships – and a reference point for many discussions of this topic – occurred at the Panguna mine of the Australian company CRA (in which UK-based RTZ had a dominant shareholding) on the island of Bougainville in Papua New Guinea. The deal to develop the mine had been struck in the late 1960s between the company and the government of PNG in Port Moresby. At the time, PNG was under UN trusteeship, administered by Australia. However, the development of the mine had the effect of creating a stark division on the island between those who benefited from the mine and those who did not. It was also claimed that the mine had done extensive damage to the ecosystems on which the local communities depended. This included pollution of the Jaba River. The situation was complicated further by the fact that many of the inhabitants of Bougainville, who are ethnically closer to those of the Solomon Islands than to those of mainland PNG, questioned the authority of the Port Moresby government to make decisions about the island and its development. In any event, to cut a very long story brutally short, there was an armed uprising against the mine, which resulted in it being closed in 1989.

If this was a particularly extreme example of a breakdown in company–community relationships, there were numerous other instances around the world where the mining industry was having to address more seriously than in the past the views of those living in the vicinity of its mines. Because mines are often found in remote regions, this often meant talking to, and reaching accommodation with, indigenous peoples. In Australia, significant efforts were being made to compensate for the earlier cavalier attitude of the mining industry towards Aboriginal people. A similar effort was under way in Canada with respect to the Inuit people. A gold rush in the Amazon region of Brazil during the 1980s threw the spotlight on the impacts of the industry on the region's indigenous peoples, notably the Yanomami. While in some cases the companies' responses to these matters were driven by the emerging reality that these local groups, if unhappy, had the capacity to close down mining operations, there was also an acceptance that the old ways of doing business, and relying on the central government to deal with local opposition to mining activity, were simply no longer acceptable.

Yet another area in which the mining industry was being challenged was with respect to its contribution to national economic development. Many of those in the industry regarded it as axiomatic that investment in mining brought positive economic benefits for the country in which the investment was made. Indeed, for low-income countries, with no significant manufacturing sector, it was possibly the only base from which their economic and industrial development could be launched. Exports of mined products would generate hard currency for a country, and mining revenues would generate tax receipts for the government. Other benefits would flow into the local economy through the spending of employees and through the cultivation and transfer of technical skills.

However, studies on individual mining countries suggested that reality was rather more complex. Some mines in emerging countries were effectively cut off from the local economy, exporting to, and supplied from, other countries. Worse, there was evidence that foisting large capital-intensive projects onto undeveloped economies could result in serious distortions, which were as likely to damage as to benefit them. The wealth generated by mining created opportunities for elites to enrich themselves but few for the poor. Inequalities resulting from such disparities of opportunity were socially divisive and encouraged corruption. It was also apparent that the impacts of increased export revenues could be double edged. Not only could these revenues be channelled into the pockets of those who preferred to spend on imported consumer goods rather than use the revenues for national economic development, but there was much evidence that the boost to exchange rates which commonly accompanied an increase in foreign exchange earnings could have the effect of raising the costs of other sectors, such as agriculture and manufacturing, rendering them internationally less competitive.

These ideas had been around for some years and their collective impacts characterised as the 'resource curse'. However, in 1995, an influential research paper by economists Jeffrey Sachs and Andrew Warner of the Harvard Institute for International Development significantly raised the stakes in the debate. Based on a statistical study of 97 emerging countries, the authors claimed that there was a 'statistically significant, inverse, and robust association between natural resource intensity and growth over the past twenty years'.[20] In other words, they claimed to have found clear evidence that resource-rich economies grew more slowly than those without natural resources.

Launching the fightback

The range and seriousness of these claims and concerns, coming on top of the parlous economic state of mining, were forcing industry leaders into a major rethink about the business they were in and their future strategies for survival and growth. Traditionally, the industry had adopted a fairly defensive posture with respect to public criticism, claiming that the industry was much misunderstood and that a lot of the criticism arose from ignorance. Certainly, it appeared to be the case that mining only ever seemed to make the news when something went wrong. The industry generally took the view that since people transparently enjoyed the benefits of mineral products in their material lifestyles, they ought therefore to accept the unavoidable consequences of digging them up. In short, the end should justify the means.

It was becoming apparent, however, that the legitimacy of mining had, in the public mind, less and less to do with the utility of *what* it produced and more and more to do with *how* it produced. It was equally becoming apparent that many of the challenges faced by the industry were interconnected. To attract money from investors, companies in the industry needed to be profitable. In order to be profitable, they needed to manage better the operational risks associated with poor environmental performance and bad community relations. In order to gain access to the best mineral deposits, they had to persuade host countries that they understood the importance of operating in an environmentally and socially sensitive fashion and that they were prepared to work with governments of mineral host countries to ensure that mining was done in a way that served the interests of national economic development. They needed, in short, to earn a 'licence to operate' in mineral host countries that went well beyond their formal legal obligations.

This insight that socially and environmentally responsible behaviour could, and should, be part and parcel of good business practice was not exclusive to the mining sector. During the run-up to the Earth Summit in Rio de Janeiro in 1992, a group of senior businessman had begun to talk about the need for industry to embrace the notion of 'sustainable development', an initiative that led to the formation of the World Business Council for Sustainable Development (WBCSD). The concept of sustainable development had its origins in the report of the 1987 World Commission on Environment and Development – usually called the Brundtland Commission after its chairperson, Gro Harlem Brundtland, a former prime minister of Norway. Although subject to numerous definitions, in its essence it sought to capture the idea that development today should be undertaken in such a manner as not to compromise the potential for development in the future.

The case mounted by this small group of businessmen, operating under the leadership of Stephan Schmidheiny, chairman of Swiss company Anova, was that, as an integral part of society and as an institution with a strong vested interest in a healthy and balanced social order, business had an important part to play in promoting sustainable development. In order to do this, it needed to define its ambitions more broadly than in the past. In addition to serving their customers and investors, businesses needed to engage constructively with other elements in society affected by their activities, be these local communities, indigenous peoples, NGOs or governmental institutions, and to operate in ways that did not result in lasting damage to our planet. It was not a question of throwing out the profit motive, but it did imply that good business performance and good environmental and social performance needed in future to go hand in hand.[21]

It was against this background that the heads of several of the world's largest mining companies came together in the margins of the World Economic Forum in Davos in January 1999 under the chairmanship of Rio Tinto's dry and cerebral chairman, Sir Robert Wilson, to discuss how they might respond to the multiple challenges facing them and how in effect they might get beyond the defensive mindset. By the Davos meeting in 2000, this had crystallised into a plan to launch a Global Mining Initiative (GMI). The declared aim of this initiative was 'to ensure that an industry that is essential to the well-being of a changing world is responsive to global needs and challenges'.[22] At its heart was to be a project, called Mining, Minerals and Sustainable Development (MMSD), which evaluated the industry in all its aspects, from exploration to recycling. The premise of the exercise was that the problems the industry faced were not simply problems of perception which could be addressed by education or improved public relations. They were fundamental problems of performance, which could only be addressed by companies demonstrating a commitment to change their behaviour. As Sir Robert put it, 'There is a gap, which sometimes feels like a gulf, between the industry's self-perception and how others see it.'[23] The industry was simply not good enough at what it was doing and was not organised properly to meet the challenges of the coming era.

As a step towards tackling this shortcoming, the MMSD was to provide a thoroughgoing review of the industry, addressing the impact of its activities on poverty, human rights, corruption, indigenous communities, biodiversity, energy use and resource efficiency. Part of the process involved the commissioning of a series of studies on different aspects of the industry and its impacts. Another part was a widespread consultation – involving a series of regional, national and international workshops – with groups having an interest in the sector, to seek out their views on where the industry was falling short and what it needed to do to play its part in a more sustainable world. From this, it was to draw up an agenda for change for the industry's consideration. Although largely funded by the mining industry, the exercise was to be conducted at arm's length. The WBCSD was to act as the contracting party with the principal researcher and coordinator of the project, the International Institute for Environment and Development (IIED), and to appoint an independent Assurance Group to guarantee the quality of the work done.

For an industry engaged in the exploitation of non-renewable resources, the application of the concept of sustainable development posed some highly specific issues. Since mineral deposits are depleted as a result of production, they cannot logically be sustained indefinitely. The focus of sustainability in the sector therefore has to be on the use that is made of the wealth generated by mining activities. In economic terms, the natural capital embodied in mineral resources does not have to be used to create new mines but can be converted into improved human capital (through education) and built capital (e.g., manufacturing plant). To the extent that this is successfully achieved, the benefits of a mine can continue to be felt long after the mine has closed. However, for this to occur, the process of conversion needs to be clearly understood and carefully managed.

The culmination of the MMSD project was a landmark report, appropriately entitled *Breaking New Ground*, which was presented to the closing conference of the GMI in Toronto in May 2002. The conference was attended by all parties to the consultation process, supporters of mining and critics alike. The report detailed what the mining industry needed to do to make a full contribution to a sustainable future. Proposals ranged over many areas, including the need to adopt best environmental practices, improved mechanisms for community engagement, the promulgation and adoption of industry standards and protocols, a greater sensitivity to, and understanding of, the development needs of host countries, a more integrated approach to mineral use and the need for better (more transparent) reporting. In addition, the report pointed up the important role that governments had to play in creating clear, effective and corruption-free conditions in which the industry could operate. At the end of the conference, the industry launched the International Council on Mining and Metals (ICMM), an industry body designed to take forward the sustainable development agenda on a continuing basis.

As the mining industry embarked on life in the twenty-first century, it was in a weak condition. Markets remained chronically depressed, and the industry remained unloved by investors and much of the public. It was, however, beginning to fight back. Through the cutbacks and consolidations of the late 1990s, the industry had rationalised its productive capacity and massively improved the productivity of its operations. It was using its capital more efficiently than ever before and was equipped to survive in the low-growth environment it assumed would persist. Moreover, it had recognised and risen to the challenge posed to its public legitimacy by embracing the idea of sustainable development. It had come a long way in putting its house in order, and it had direction. Everything, however, was about to change.

2
China Changes Everything

Where did that come from?

During a brain-storming session at a major iron ore producing company in 2000 executives were trying to get a handle on what might be expected of China ten years hence. At that time, China was producing around 125 million tonnes of steel a year. After torturing the data for several hours over numerous cups of coffee, the consensus of the meeting was that steel production in China would probably rise to around 180 million tonnes a year before reaching a plateau. At the outside, it might reach 200 million tonnes. In the event, China produced 639 million tonnes of steel in 2010. The forecasters had been out by a factor of more than three.

The fact that this group of executives should have got things so spectacularly wrong is testament to the hazards of forecasting. But it is also understandable. No country had ever come close to these levels of steel production before. Japan's production had peaked at 119 million tonnes in 1973 and that of the United States at 137 million tonnes in 1974. China's 2010 output was completely off the scale of historical experience. The forecasting error served to emphasise how poorly Western companies understood China. For many, the country was still pretty much a black box.

By 2000, the period of economic reform in China initiated under the pragmatic leadership of Deng Xiaoping had been in train for two decades. Even so, China's impact on the global economy and on global commodity markets had remained modest. In fact, somewhat ironically in the light of what subsequently occurred, China was better known to the rest of the world as an exporter rather than an importer of mineral commodities. Among the products it exported were specialist minor metals such as antimony and tungsten and so-called industrial minerals such as fluorspar, magnesite and talc.

China's published statistics showed that its economy had been growing at an average of just over 9 per cent a year through the 1980s and just over 10 per cent through the 1990s. However, fast as it may have been growing, China was coming off a very low base and the country's impact on the global economy was limited. It lacked critical mass. The same applied to its impact on global mineral markets. Although China's demand for minerals and metals was growing rapidly, its intensity of use of minerals was generally flat and its draw on mineral markets still small.

However, even if not very visible to Western eyes, China was in the 1990s laying the foundations for what was about to occur. The decade had started strongly, rather too strongly as it turned out. Gross domestic product (GDP) growth over 1991–1994 averaged almost 14 per cent a year, giving rise to acute inflationary pressures. Consumer price inflation surged to an average of 24 per cent in 1994. In July 1993, China's economic strongman, Vice-Premier Zhu Rongji, launched an austerity programme to contain inflation, restrict credit growth and clamp down on fraud and corruption. By 1997, these policies had restored China's economy to better balance and the administration was ready to move on.

It had become apparent that China's institutional structures, with their emphasis on state ownership and administered pricing, were becoming an obstacle to future economic growth. The large state-owned enterprises (SOEs), which were the backbone of the Communist economy and which still accounted for half of all China's industrial production, were bloated, inefficient, dirty and heavily indebted. The banks were in turn being dragged down by these companies and, because of the scale of their non-performing loans, were being prevented from directing capital to new, more profitable businesses.

In 1997, at the 15th Party Congress, Jiang Zemin, the president, and Zhu Rongji, now premier designate, launched a programme of radical economic reform. This was to involve cutting back and restructuring the SOEs, recapitalising the banking sector, rationalising government departments and liberalising internal markets. The objective was a more efficient economic system giving greater emphasis to profitability.

To mitigate the negative effects of industrial reform on employment, the government at the same time committed itself to a massive programme of public spending. This was to be directed in particular to infrastructure developments in rural areas and to the revitalisation of regions affected by SOE cutbacks. The government sought to cement the shift to more liberal economics by opening up the Chinese economy to global markets by pursuing membership of the World Trade Organization (WTO).

Although not so apparent at the time, this marked the beginning of a completely transformational period of growth for China, with far-reaching implications for the mining industry. These implications stemmed not so much from the acceleration of China's growth but from changes in the nature of that growth which rendered its economy more material intensive. The remainder of this chapter addresses the drivers of China's material-intensive growth and the effect this had on individual commodities, and considers the implications of recent economic developments.

Drivers of growth

On the west bank of the Huangpo River in Shanghai stands The Bund, a collection of solid European-style buildings, which, in the 1920s, formed the financial centre of the Far East. At that time, and indeed up until the early 1990s, the eastern bank of the Huangpo, facing The Bund, was largely fields and villages, with a few warehouses and wharves along the shoreline. In 1993, the Chinese government announced the establishment in this area of a special economic zone. This was to incorporate the Pudong New Area, which it determined would become the financial centre of the new China. Over the next few years, Pudong became one enormous building site as an array of gleaming skyscrapers shot up and transport links to the area were laid down. The Oriental Pearl Tower, until 2007 the tallest building in China, became the instantly recognisable symbol of the project and of its ambition. Looking around the site at the turn of the millennium, visitors could not help observing how many of the buildings seemed empty and wondering whether all this construction was warranted. Not long after, they had their answer; the place was full. Today, Shanghai Pudong is a city of 5.5 million people. China's policy of 'build it and they will come' had been set in motion and was working.

The period of transformational growth on which China was now embarked had three core, interconnected elements: exports, investment and urbanisation.

Exports

With its large population and widespread poverty, a central objective of the Chinese leadership was the maintenance of social cohesion through the delivery of economic growth. As the ideological basis of China's governance withered, the delivery of economic growth became the foundation of China Communist Party (CCP) legitimacy and the grounds on which it could justify its robust economic policies and authoritarian politics. Early on, the leadership of the new China had determined that trade was a key to achieving rapid growth. China simply did not yet have the domestic market to drive growth internally. With respect to trade, China had one particular advantage. While it might not have the benefit of advanced technology or abundant natural resources, it undeniably had a lot of people. Historically, large poor populations had tended to be viewed as an obstacle to growth rather than an opportunity. But in the eyes of the Chinese leadership a lot of poor people equated to a lot of low-cost labour. China could use this to compete its way into global manufacturing markets.

In choosing to go this route, China was in many respects simply following the successful export-led model adopted before it by other Asian countries, including Japan, South Korea and Taiwan. However, there was one crucial difference. While these other countries had generally used internal corporate and financial resources to build their export sectors, China decided to expedite matters by throwing open its doors to foreign investors. Western companies jumped at the opportunity, quickly transferring the low-wage parts of their supply chains to China in order to reduce their costs of production – a process referred to as 'offshoring'. Foreign direct investment (FDI) into China surged from $3.5 billion in 1990 to $36 billion in 1995, by which time China was accounting for fully one-third of FDI going to emerging countries. These funds continued to flow in for the next decade as companies around the world moved their manufacturing activities to China. In 2005, US retailer Walmart imported goods valued at $18 billion a year from China and was China's eighth largest 'trading partner'.[1]

A major boost to China's trade growth was provided by its entry to the WTO at the end of 2001. China had been in discussions about signing up to the trade regime, the General Agreement on Tariffs and Trade (GATT), which the WTO operates, for many years but the brutal suppression of the Tiananmen Square protests in June 1989 had put these talks on hold. However, for President Jiang Zemin and his reforming premier, the matter of China's entry into the WTO had acquired a new urgency. Not only were they anxious to see the economic benefits that membership would bring but the trade liberalisation measures which WTO membership required also gave the leadership a handy justification for the aggressive pursuit of their reform agenda. Outside the WTO, China's trade relations were a ramshackle collection of bilateral arrangements. Inside, its exports had the same access to foreign markets as those from other member countries and the same protections from the imposition of arbitrary restrictions.

In 2002, as the world economy began to emerge from the effects of the 2000 dotcom bust, China's export growth found another gear. After growing 7 per cent in 2001, China's exports (expressed in US dollars) expanded 22 per cent in 2002 and 35 per cent in 2003. Global industrial production (IP) was relocating to China to take advantage of low production costs, and the country's factories were pumping out electronic goods, domestic appliances, car components and textiles for consumers around the globe. China was becoming the workshop of the world. The close relationship between China's export growth and the acceleration of its industrial production during these years is illustrated in Figure 2.1. Growth in IP rose from a rate of 10 per cent in 2001 to 17 per cent in 2006 and 2007 before subsiding under the impact of the global financial crisis (GFC). That the rates of growth of industrial production exceeded the overall growth of China's economy in these years (when GDP averaged around 10 per cent a year) revealed something about the changing structure of the Chinese economy and its increasing material intensity. Exports were growing as a component of GDP, and proportionately more of the economy was being devoted to the production of things rather than to agricultural production or the provision of services. The surge

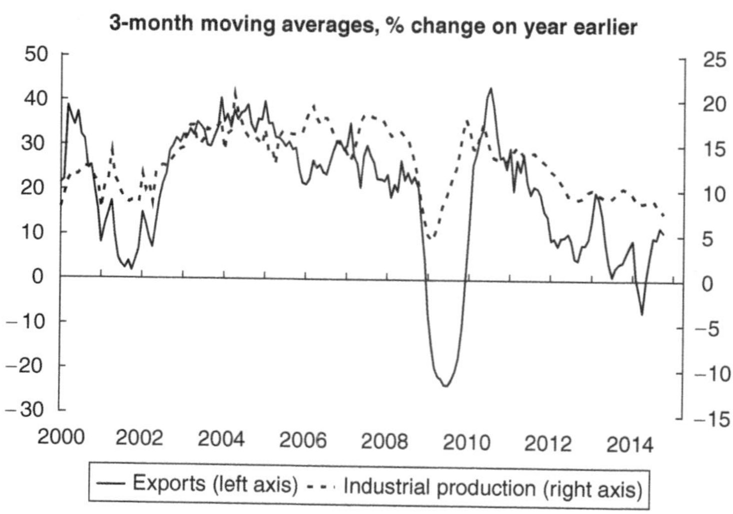

Figure 2.1 Growth in China's exports and industrial production, 2000–2014
Source: China Customs and NBS.

in China's demand for mineral raw materials coincided with, and was substantially driven by, this explosion of exports.

Investment

The second component of China's extraordinary growth during these years was investment. To equip itself to become the workshop of the world, China needed to invest heavily. In addition to building and equipping factories, China needed to build the power stations which supplied their electricity, the railways and ports which brought in their raw materials and the ships and containers which freighted their products out. Because earlier regimes had emphasised local self-sufficiency, relatively little effort had been devoted to connecting together China's vast, mountainous and river-strewn interior. In short, it needed a lot of infrastructure. China also needed to build accommodation for the workers drawn to the cities along with the shops and offices to service their requirements. And, of course, it needed to construct the plants to furnish the raw materials for all this investment activity.

The trends in investment are reflected in Chinese data on fixed asset investment (FAI). These data have their limitations. They include certain property transactions and they are not adjusted for inflation. But there is no reason to doubt the general picture they present. Between 1996 and 2000, data from the National Bureau of Statistics (NBS) show FAI growing at an average rate of under 9 per cent a year. In 2001, this accelerated to 15 per cent and in 2002 to 18 per cent. In 2004, it grew at a breathtaking 38 per cent. While this was clearly unsustainable, growth in FAI nonetheless remained at an elevated rate of 25–30 per cent for the remainder of the decade. All the three principal components of FAI – manufacturing, real estate and infrastructure investment – contributed to this growth, although, through the middle part of the decade, manufacturing investment grew the fastest.

Funding this surge of investment was not a problem. In common with many other countries in Asia, China has a very high savings rate, a fact attributable in part to the lack of welfare provision and the consequent need for people to put money aside for their retirement and for health care and education. Without the ability to move their money out of China, these savings are effectively captive to the banking system, providing China with a large reservoir of cheap and ready finance. Data from the World Bank show the savings rate of the Chinese population rising from an average of about 40 per cent during the 1990s to over 50 per cent in the decade beginning 2002.[2]

There are strong inbuilt incentives in China to ensure that this high level of savings is matched by high levels of investment. The SOEs in China have a close symbiotic relationship with the CCP. The SOEs provide political support and personnel to the Party while the Party provides funding for the SOEs, this notwithstanding that the state sector has much lower productivity than the private sector. The nature of the relationship between central and local government in China also provides structural support for investment. The performance of local officials has traditionally been assessed on their ability to generate growth and to provide jobs. Accordingly, many local governments work closely with banks and developers to bring forward and build projects in their regions, not infrequently riding roughshod over local landowners and environmental concerns in the process. Although the Chinese leadership is obliged to maintain the fiction that the economy is centrally directed and that it (the government) is pulling the levers, the reality is that a lot of investment is driven from a local level and is loosely regulated. The ability of local authorities to operate without much direction or interference from Beijing is neatly captured by the Chinese aphorism, 'The sky is high and the emperor is far away'.

For all of these reasons, China's investment was growing rapidly, not just in absolute terms but also as a share of GDP. Investment as a component of GDP is referred to as gross fixed capital formation (GFCF) and is statistically distinct from FAI. GFCF typically accounted for 30–35 per cent of GDP until 1992 when it shifted into a range of 35–40 per cent (Figure 2.2). While high, these levels were not wholly out of line with those experienced by other Asian countries undergoing rapid industrialisation. What happened next, however, took it into new territory. As China geared up for its export surge, investment rose through 40 per cent of GDP in 2003 and by the end of the decade had risen to 48 per cent. As can be seen in Figure 2.2, this growth in investment came largely at the expense of private consumption, the share of which declined from 46 per cent in 2000 to 35 per cent in 2010.

Urbanisation

The third element driving China's growth over this period was urbanisation. The issue of urbanisation is often talked about as if it were an independent driver of economic development. In reality, urbanisation was closely linked to the other two components of China's industrialisation, export growth and investment. Rising exports and industrial production drew people into the cities, while the movement

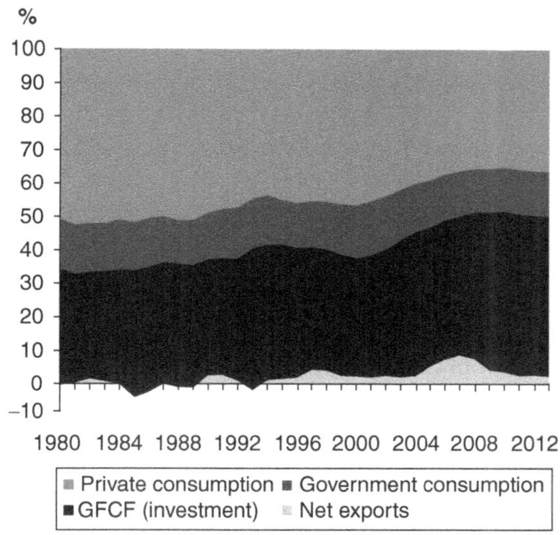

Figure 2.2 China: Composition of GDP, 1980–2013
Source: NBS.

of large numbers of people to the cities in turn generated demand for investment in housing and transportation.

The shift of population from low-wage agricultural occupations in the countryside to higher-productivity jobs in the cities is a key feature of the process of industrialisation everywhere. Accordingly, it was a central plank of Deng Xiaoping's plan to raise China's living standards. And it was a key element in the Zhu Rongji reforms of the 1990s.

At the time of Deng Xiaoping's elevation to the leadership of the CCP, the urban population of China stood at around 18 per cent of the total and had been stable for years. Over the following years, the share of the urban population edged up gradually to reach 30 per cent by 1995. In 2013, urban Chinese represented 50 per cent of the country's total population of 1354 million. In terms of actual people, this meant that there were 345 million more people living in the cities in 2013 than in 1995. This is more people than live in the United States.

These trends are important from the perspective of mineral markets since the economic impacts of urban dwellers are different from those of their rural counterparts. Not only do they tend to be more economically productive but they use a lot more resources. This is partly because their housing, transportation and services such as power and water supplies use more resources, but it is also because the urban middle classes, with higher levels of disposable income, greatly increase the demand for consumer durables such as vehicles and electrical appliances. Above a certain income (roughly $5000 per head), discretionary spending on consumer durables increases disproportionately fast. It is estimated that a typical Chinese urban household accounts for 10–15 times more steel use than a rural household.[3]

There was also an important geographical dimension to China's process of urbanisation. Under the regime of Mao Zedong, the focus of industry within China had been in the north and the north-east, in proximity to the large Shanxi coalfields. Moreover, since Mao Zedong's model of development emphasised small-scale and internal development, much of China's industrial production during this era took place at locations away from the coast.

The refocusing of the Chinese economy towards exports required that the new push for industrialisation took place on the coastal plains of China to the east and south of the country. This implied large amounts of new investment. Factories sprang up along the Pearl River Delta in the province of Guangdong and along the Yangtze River Delta in Zhejiang, Jiangsu and Shanghai. Moreover, infrastructure was required to service these factories and to connect China to its suppliers of raw materials and its market in other countries. And as the investment poured in the population of China flowed eastwards from the rural economy in the interior to build and to work in the new and growing coastal cities.

There is a particularity about building in China's cities that adds to their material intensity. China has an acute shortage of good agricultural land and much of what it has is to be found on the coastal plains, which is to say the same areas in which China's industry is concentrated. So as not to build on more of its agricultural land than is absolutely necessary, there is a strong imperative to construct high buildings. In 2000, around 60 per cent of the residential floor space in the cities was in buildings seven stories or more high. By 2010, this had risen to 70 per cent and by 2020 is expected to be 90 per cent, with some 40 per cent being in buildings of 16 or more stories. As buildings get taller, their use of steel increases disproportionately. Thus, while a building of 8–12 stories typically employs around 50 kilograms of steel per square metre of floor space, residential buildings of 35–50 stories require roughly twice as much steel per square metre.[4]

As a consequence of these three interconnected forces – export growth, investment-driven growth and urbanisation – China achieved a rate of industrialisation wholly unprecedented in economic history. The benefits of being able to 'leapfrog' technologically, of course, permits countries industrialising later to do so faster than those who blaze the technological trail. Even so, the speed of China's growth was remarkable. During the course of Britain's industrialisation, starting in 1780, it took 58 years to double its GDP per head. In the United States it took 47 years and in Japan 34 years. The Republic of Korea managed it in only 11 years, starting 1966.[5] China routinely doubled its GDP per head every eight years through the 1980s and 1990s and every seven years through the 2000s. No less importantly from the perspective of the story being told here, China's growth, and the specific form which this growth took, resulted in the biggest surge in the demand for mineral raw materials ever experienced in history.

Impact on individual commodities

The impact of China's growth on the global demand for minerals in the early part of the twenty-first century is hard to exaggerate. Some of the basic facts about China's demand growth and its impacts are presented in Table 2.1. For the purpose of this table, the period used is that between 1998, the year the Zhu Rongji reforms were committed, and 2008, the year of the GFC.

As the data make plain, China's consumption of most of the mineral commodities listed here rose at an extremely rapid pace over the period 1998–2008. Except in the case of the precious metals, the rates of growth were significantly higher than the growth of China's GDP, attesting to

Table 2.1 Growth in consumption of selected commodities: China and the world

	Average annual % consumption growth 1998–2008			**China consumption as % of total world in:**		
	China	**Rest of World**	**Total World**	**1998**	**2008**	**2013**
Aluminium	17.7	2.2	5.3	11	36	48
Copper	13.9	1.0	3.1	11	28	47
Nickel	23.2	0.5	3.0	4	26	51
Zinc	13.9	0.7	3.8	14	36	46
Steel	14.9	3.0	5.9	16	37	47
Iron ore*	23.9	1.1	7.4	13	52	65
Gold+	4.5	−2.8	−2.1	7	13	31
Platinum+	7.1	2.9	3.5	12	18	28

Note: *Consumption of seaborne iron ore. + Fabrication demand.
Source: Author's calculations, based on GFMS, Johnson Matthey, UNCTAD, WBMS and WSA.

the increased material intensity of China's economy over this period. The data also reveal how slow was consumption growth in the rest of the world and how much of the growth in global consumption was therefore attributable to China. Finally, the table shows how China's share in global consumption grew over this period, turning China from a modest consumer, accounting for 10 per cent or so of global consumption in 1998, to a dominant player in global markets, accounting for a quarter to a third or more of global consumption in 2008. As is also apparent from the table, this dominance of China grew still further after the GFC. By 2013, China accounted for approaching 50 per cent of global consumption of the industrial metals. In fact, over this longer period, 1998–2013, China effectively accounted for the entirety of global growth in consumption for many commodities, including copper, nickel, seaborne iron ore and gold. This represented a degree of dominance in minerals use unmatched since the early 1950s when the United States similarly dominated world metal markets. It was, however, a dominance somewhat out of line with other indicators of economic importance. China's share of global GDP (on a PPP basis) in 2013 was, according to the IMF, 16 per cent, and its share of global industrial production around 19 per cent. Its share of global oil use was 12 per cent.[6]

Although China's growth in demand impacted all markets, the degree of impact was variable between commodities. The biggest relative impacts were experienced by steel, iron ore and aluminium. Global demand for these commodities had, over the period 1980–2000, grown at roughly 2 per cent a year. However, as Table 2.1 illustrates, over the period 1998–2008, growth in global demand for steel accelerated to 6 per cent, that of iron ore to 7 per cent and that of aluminium to 5 per cent. In other words, China's growth in demand for these commodities represented a substantial addition to the global total and resulted in a rapid acceleration in demand at a global level.

The situation for copper and nickel was slightly different. Although China undoubtedly contributed towards an acceleration of global demand, the impact was more limited; effectively an increase from an annual growth of 2.4 per cent to 3 per cent for the two metals between the periods 1980–2000 and 1998–2008. This suggests that a significant part of the growth in China was coming from the displacement of demand in other regions rather than being additional to it. In other words, it was related as much to the relocation of global industrial production to China, and to the associated shift in trade patterns, as it was to demand arising from within the Chinese economy itself. China's impact on world markets in the case of gold and platinum was more muted still. Growth in demand for gold in China over 1998–2008 failed to prevent a decline in world use of gold for purposes of fabrication. In the case of platinum, global demand for fabrication decelerated from 6 per cent a year over 1980–2000 to 4 per cent over 1998–2008, although the analysis of demand trends for this metal is complicated by supply constraints during this latter period.

So, what was China using these mineral commodities for?

The use of steel – and thus, by implication, the iron ore used to produce steel – is heavily focused on investment activity and in particular construction (Figure 2.3). Something of the order of 55 per cent of steel in China goes to building commercial and residential properties and to creating infrastructure, for example, power stations, ports and railways. Accordingly, steel use tends to track FAI quite closely. The second biggest use of steel is in the fabrication of machinery, another investment item, which accounts for around 15 per cent of total steel use. The balancing 30 per cent is accounted for by a variety of applications, including auto production, shipbuilding and the manufacture of consumer appliances.

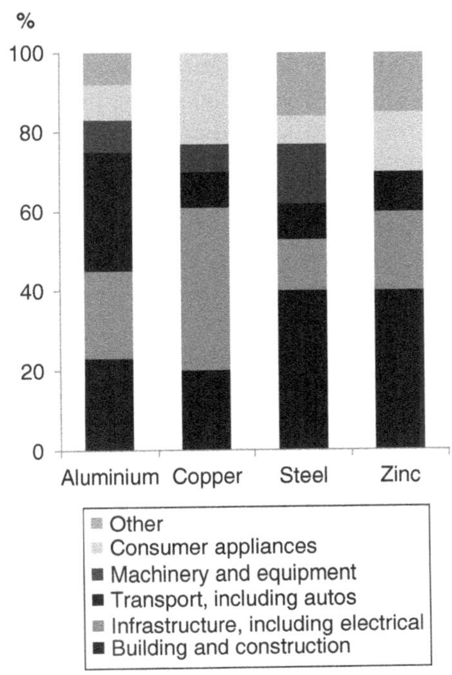

Figure 2.3 Use of metals in China
Source: Macquarie Capital (2014).

Construction is also a key factor accounting for the rapid growth in the use of zinc and aluminium. Most zinc goes into protective coatings for steel (zinc galvanising), thus its use closely matches that for steel. In the case of aluminium, around a quarter of the metal's use is accounted for directly by construction. The other major uses of aluminium in China are also strongly focused on investment products. In order of importance, these are transportation goods, electrical applications (e.g., high tension power lines) and machinery. By comparison with what occurs elsewhere in the world, China uses relatively less aluminium in consumer packaging (cans and foil).

Copper use in China is, as elsewhere, dominated by electrical applications. Around 40 per cent of copper in China goes to power infrastructure; that is, to power stations and power cables. A further 20 per cent or so goes to construction, which includes building wire. The remaining uses of copper are dominated by consumer products. These include wiring and contacts for electrical and electronic products, such as home appliances and computers, but also the thin winding wire used to make electrical motors. They also include copper use in cars and copper tube used in the production of air conditioning and refrigeration units, in both of which China has a major export trade.

By comparison with some of the more basic materials of industrialisation such as steel and cement, nickel is always something of a late runner in the race for industrial development, its demand being driven more by lifestyle products. This explains why its use in China was so low relative to that of the other metals in 1998. However, when China did start to use nickel seriously, its demand expanded very rapidly, as Table 2.1 shows. Much of nickel's use is accounted for by the production of special steels, in particular stainless steels, which are employed in relatively sophisticated products such as high-end consumer goods (e.g., top-of-the-range kitchens) and equipment for use in the food industry, hospitals and petrochemical processing. Stainless steel facings for buildings are also popular in China as are decorative nickel coatings.

The precious metals gold and platinum are, as is the case with nickel, late runners, their principal use in China being in jewellery, demand for which is driven by disposable income.

China and the rest of the emerging world

China's rapid industrialisation in the first decade of the twenty-first century fitted into a broader narrative about how the balance of the global economy was shifting towards emerging market economies.

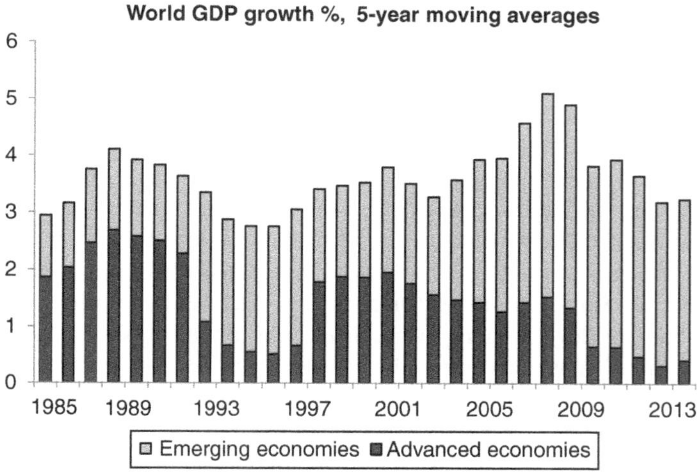

Figure 2.4 Composition of global growth, 1985–2013
Source: Based on IMF.

Strong growth in China and in other emerging economies over many years meant that the historical dominance of the global economy by the advanced Western economies was being eroded. As Figure 2.4 illustrates, through the 1980s and 1990s, global growth was still largely dominated by the advanced economies. In the 2000s, growth rates (based on PPPs) moved higher and growth was dominated by the emerging economies. According to the IMF, the emerging markets' share of global GDP, which was 37 per cent in 2000, crossed the 50 per cent mark in 2013.[7]

A landmark study by economists at the US investment Goldman Sachs in 2003 helped both to dramatise and popularise this transition.[8] The focus of the study was on four emerging market economies – Brazil, Russia, India and China – which Goldman Sachs christened the 'BRICs'. Using a simple economic model employing working population, capital stock and technical change (estimated by assuming that productivity in emerging economies gradually converges on that of advanced economies), the Goldman Sachs team forecast that by the year 2025 the BRIC countries would have a combined GDP half that of the G6 (the United States, Japan, Germany, France, Italy and the United Kingdom) and that by 2040 its combined GDP would exceed that of the G6. At the time the report was written, the proportion was just 13 per cent. (Note that the GDP data used in this study were based on prevailing exchange rates, not PPPs.)

The transition described by this analysis was in part underpinned by the ongoing globalisation of industrial supply chains. Notwithstanding the convergence of emerging and advanced economies, most of the world's disposable income continued to reside in the more advanced economies, the United States, Western Europe and Japan. These economies were generally keen to retain control of the financing, design, and marketing of the products sold in their markets. But with respect to the manufacturing of these products, they were increasingly taking advantage of the lower labour costs offered by emerging economies. Some service sector activities, such as back office services and software development, were also being moved offshore. With the migration of manufacturing, the consumption of mineral raw materials was also migrating. Already, by 2000, emerging economies were consuming over 50 per cent of the world's steel.

In light of these trends and of the forecasts that the shift in economic power towards the emerging economies had a long way still to run, analysts began to speculate that something of major long-term significance was taking place in commodity markets. Not only were emerging economies growing faster than the advanced economies but, reflecting their stage of industrial development, their growth was more material intensive. Accordingly, some began to hypothesise that the world stood on the threshold of a multi-decade period of material intensive growth, allowing the mining industry decisively to put behind it the painful experiences of the 1980s and 1990s. Thus was born the notion of a commodities 'super cycle'. This is discussed in more detail in Chapter 4.

Although the growing importance of emerging economies was broadly based, from the perspective of raw materials demand the story was essentially about China. In absolute terms, China's economy was bigger than the other three BRIC economies put together. Moreover, among the BRICs countries, the importance of Brazil and Russia arose more from their role as raw materials' producers rather than consumers. Their consumption of mineral raw materials was a small fraction of that of China. India, although projected by Goldman Sachs to outpace all the other BRICs countries in the later years covered by their study because of its supportive demographics, was following a markedly different growth model to China, and its consumption of mineral commodities was only some 10–15 per cent that of China's. Certainly, there was the potential for India, and indeed for other populous emerging economies such as Indonesia, Mexico, Turkey and Nigeria, to see a sharp increase in their use of mineral raw materials in the course of time but, for the moment, China was overwhelmingly the dominant player.

Impact of the Global Financial Crisis

Wenzhou, in the south-eastern province of Zhejiang, is symbolic of China's economic revolution. Its spirit of enterprise (it was the crucible of China's private sector in the early 1980s), combined with its coastal location and supply of low-cost labour, led to the creation of numerous successful export-oriented companies. In a city of a little over nine million people, there are estimated to be around 450,000 businesses. These produce most of the world spectacles and cigarette lighters and are also famous for shoes and electrical goods. In the immediate aftermath of the 2008 GFC, Wenzhou initially bore up better than most other Chinese cities. However, as time passed, cracks began to appear. Softening export markets and rising costs began to depress sales and cut into profit margins. Attention among Wenzhou's entrepreneurs began to shift to the use of cheap credit to speculate in financial markets and in property. Stories began to emerge of dozens of factory owners absconding, leaving workers unpaid and mountains of debt. In view of the city's symbolic significance, in October 2011, Premier Wen visited the city to inform himself personally about what was going on and to pledge government support for struggling businesses. Although Wenzhou subsequently recovered its poise, its days of high-energy, breakneck growth, based on extreme business risk and mass immigration, appeared to be over.[9] Its experience serves as a microcosm of what went on in China during these years.

The financial crisis into which the world was plunged in the second half of 2008 following the collapse of Lehman Brothers posed a major challenge for China and badly shook confidence in the market for commodities. The credit binge in the advanced economies, which had fuelled their consumption-led growth and provided a booming market for China's exports, came to an abrupt end. Global GDP shrank for the first time in the post–World War II era. China's exports fell by a third between the second half of 2008 and the first half of 2009. Early in 2009, it was reported that some 20 million workers had been laid off in China. Commodity prices dropped like a stone. The copper price, which had stood at an average of $8684/tonne ($3.94/lb) in April 2008, had fallen to $3071/tonne ($1.39/lb) by December.

Given the dependence of China's growth on a strong export performance (20 per cent of GDP growth between 2004 and 2007 was accounted for directly by net exports, in addition to that attributable to export-related investment), there had been speculation among economic analysts for some time about how robust China's growth would prove to be in the absence of an external stimulus. The debate had commonly been cast in terms of whether China's growth could be 'decoupled' from that elsewhere. The matter was about to be tested and the early signs were not encouraging. China's growth in industrial production in the last quarter of 2008 dropped to its lowest level in six years. Shanghai's stock market fell sharply while property sales dropped 40–50 per cent below those of a year earlier.

China's response to the collapse in exports was to announce, through the State Council in November 2008, a four trillion renminbi ($586 billion) stimulus programme. This was the largest of any country relative to the size of its economy, equating to roughly 14 per cent of GDP and in dollar terms four times the size of the US fiscal stimulus announced earlier the same year.[10] The focus of the stimulus package was to be public investment projects in transport (especially railways and airports), rural infrastructure (water supply, irrigation, rural roads and power grids), environmental protection (sewerage and green energy) and public housing. Subsidies were offered to those in rural areas to buy consumer appliances such as refrigerators and televisions in order keep factories running. In support of these projects, the government announced it would loosen credit and encourage state banks to lend.

The impact of the programme was swift and dramatic. In remarkably short order, increased domestic investment took up the slack left by the slowdown in exports, manufacturing and real estate. Transport sector FAI in the first half of 2009 ran at over 60 per cent above year-ago levels. Growth in GDP, having dropped in the first quarter of 2009 to a rate of 6.2 per cent, by the first quarter of 2010 was back up at 11.9 per cent. The rate of growth of industrial production over the same period rebounded from 5.1 per cent to 19.6 per cent.

As the Chinese economy came storming back, so too did its demand for commodities. Many observers of commodity markets had in early 2009 reluctantly accepted that the boom was probably over and that prices quite probably faced a long hard slog out of the trough. This had been the case with previous experiences of recession, such as those in the early 1980s and early 1990s when global demand for commodities had fallen sharply. And it was certainly the case that most advanced economies were in a bad way, with industrial production falling sharply and demand for mineral commodities falling with it.

Given the shift in the locus of global industrial production, however, China had become the most important component of world mineral commodity demand. And it was recovering fast. Although the extent of the global economic slowdown was such that even China was unable to prevent demand for most minerals declining in 2009, by the second half of that year markets were beginning to firm again. China's use of minerals and metals continued to grow in 2009 as it cemented its position as a global manufacturing power and as the state-backed investment programme sucked up resources. Chinese users of minerals were also adding to demand by taking the opportunity of lower prices to build up stocks. In addition, the government was buying for its strategic stockpile, held by the Strategic Research Bureau (SRB). During the course of the year, the SRB reportedly bought 590,000 tonnes of aluminium, 235,000 tonnes of copper and 159,000 tonnes of zinc, as well as a variety of other minor metals.[11]

China's robust economic policies ensured that it continued to provide support for the global economy and for commodity markets. Following its rebound in 2009, China's GDP grew by 10.3 per cent in 2010 and 9.3 per cent in 2011. Although growth subsequently slowed, it was still substantially faster than that being experienced elsewhere. Any threats to growth were met by new spending initiatives and new injections of credit. The 12th five-year plan, covering 2011–2015, committed the government to the construction of 36 million units of low-cost housing. In September 2012, in the run-up to the change of leadership at the end of the year, the government announced the approval of 60 infrastructure projects costing $157 billion. April 2014 saw the announcement by the State Council of a mini-stimulus package comprising a further 80 new major public projects covering railways, ports, IT infrastructure and renewable energy. Credit made available through the banking system and, increasingly, the (unofficial) 'shadow' banking system, spurred rapid growth in spending on real estate.

The strength of China's recovery from the GFC ensured that its role in the global economy and its dominance of mineral markets increased still further. As Table 2.1 showed, by 2013, China was totally dominating global demand for many industrial commodities. The pressure on recession-hit manufacturers in advanced countries to cut their costs and protect their margins added to the attractions of outsourcing to China. Moreover, a significant part of the monetary stimulus in the advanced economies, far from inducing an increase in domestic economic activity as intended, was simply leaking out into further purchases from China.[12] According to the Organisation for Economic Co-operation and Development (OECD), industrial production in the G7 countries in 2013 was still 6 per cent below its level in 2007. Over the same period, China's industrial production almost doubled. In 2009, China stormed past Japan to become the largest producer of cars in the world, and by

2013, it accounted for 28 per cent of all cars produced. In 2014, the World Bank claimed that on the basis of revised estimates of purchasing power parities, China had grown to become the largest economy in the world.[13]

Growing imbalances

China's recovery from the GFC, remarkable as it was, served to expose mounting strains in its growth model and raise serious questions about the sustainability of its growth performance.

Although China's exports had continued to grow following the GFC, its saturation of export markets and the poor state of the rest of the world economy meant that exports no longer had the capacity to galvanise further growth in GDP. The relocation of global industrial production, which had driven China's growth following its accession to the WTO, and which had prompted the massive surge in its demand for minerals, could not be repeated. Moreover, the huge cost advantages enjoyed by China's manufacturers were gradually beginning to erode as real wages in China rose and those in the advanced economies fell. Rising housing costs and a strengthening currency were further contributing factors, as was a growing reluctance of the rural population to travel to the cities and suffer the privations of migrant city life. China's unlimited supply of cheap labour was proving not so unlimited after all.

With the loss of momentum from exports, the maintenance of China's growth had to rely increasingly on internal drivers of growth, particularly investment. Official figures showed that over 90 per cent of China's GDP growth in 2009 came from investment.[14] In the immediate aftermath of the GFC, the main emphasis of this investment had been infrastructure spending, but by 2010 the focus had shifted to real estate.

However, investment as a driver of growth was running up against its limits. By 2010, the share of investment in China's economy was approaching an unprecedented 50 per cent of GDP. The high level of investment was creating overcapacity across all sectors of the economy, manufacturing, infrastructure and property. Visitors to China were returning with tales of ghost cities, unoccupied office blocks, shopping malls without shoppers, empty airports and bridges to nowhere. A spokesman for the China Iron and Steel Association (CISA), Li Xinchuang, stated in early 2014 that China's steel industry had some 300 million tonnes a year of surplus capacity (around 30 per cent) and that the problem of overcapacity is 'probably beyond our imagination'.[15] Each additional unit of capital spending in the economy was buying less and less output, a trend not unrelated to the fact that the government was channelling a lot of its spending through inefficient, unproductive SOEs. Over the period 1990–2007, the incremental capital–output ratio (a measure of the amount of investment required to generate one unit of GDP) averaged 3.8. Between 2008 and 2014, this rose to 5.5.[16] A report from China's state planning body, the National Development and Reform Commission (NDRC), estimated that over the period 2009–2013 'ineffective investments' amounted to $6.8 trillion, almost half the total invested.[17]

Aside from being inefficient in its use of capital and fundamentally deflationary in its impacts, this investment-heavy model of growth had several other important characteristics. It encouraged corruption, it was highly material intensive and it was environmentally destructive.

The rapid ramping up of state spending in the wake of the GFC, and the increases in credit availability which accompanied it, created widespread opportunities for corruption among central and local government officials and resulted in the construction of substandard buildings and an increased incidence of bad debts.[18] These issues came together symbolically in a dramatic high-speed rail crash, which took place outside the city of Wenzhou in July 2011. The accident, in which 40 people died and 192 were injured, triggered a vigorous and high-profile crackdown on corruption. The rooting out of corruption was made a key priority of Xi Jinping when he assumed the presidency in 2012, and the subsequent campaign against corruption stretched from the deputy head of the People's Liberation Army down to the lowliest government official.[19]

The unusually high material intensity of China's growth model can be gauged from the fact that it uses nine times as much steel to generate $1 million worth of GDP (PPP based) as does the United States and four times as much as Japan and Germany. Even comparing China with other emerging economies, the differences are stark. It uses five times as much steel as Brazil and four times as much as India to generate the same amount of GDP. Extending the analysis across aluminium, copper, nickel and steel, and across the global economy, China consumes three times as much metal in generating $1 million of GDP as the average including China and five times as much as the average excluding China.[20]

China's investment and material-heavy form of growth was creating severe environmental problems. Industrialisation often takes a toll on the natural environment, but the form and speed of China's industrialisation meant that environmental damage was occurring on a truly heroic scale. Air pollution from industrial plant and vehicle emissions meant that air quality in many cities was multiples of the levels the World Health Organization (WHO) considered safe. A study published by the US National Academy of Sciences found that air pollution in the north of China reduced life expectancy by five and a half years.[21]

There is also a chronic water problem. China has 20 per cent of the world's population but only 7 per cent of its fresh water resources. As a consequence of the form of its agricultural and industrial development, China has been running through, and polluting, these resources at a dangerous rate. Water is underpriced and the productivity of its use is low. China uses four times more water per dollar of GDP than its G20 peers while its steel mills use four to nine times more water per tonne of steel produced than do those in Japan and the United States. Two-thirds of China's 660 cities suffer from water shortages with the situation in 110 classified as severe. In northern China, a region containing half the country's population and two-thirds of its agriculture land but which is also the driest part of the country, over 90 per cent of the aquifers under cities are polluted. Because China is using more water resources than can be replenished by rainfall, it is digging deep into its groundwater reserves and the water table across much of the north of the country is falling at 1 metre per year.[22] Concerns over air and water pollution have resulted in a growing incidence of public protests in recent years.

The leadership of China had for some time acknowledged the need to rebalance the economy towards a model giving more emphasis to consumption and to the quality of growth, which is to say growth directed more towards greater social equality and having more regard for the environment. It was intended to be a central plank of the 11th five-year plan, covering the years 2006–2010. The premier of the day, Wen Jiabao, had himself declared in 2007 that China's economy had become 'unstable, unbalanced, uncoordinated and ultimately unsustainable'.[23] The 12th five-year plan, covering 2011–2015, reiterated the government's intention to shift the emphasis of growth towards consumption and to give greater prominence to the environment, resource efficiency and social welfare. At the CCP's Third Plenum in November 2013, the new leadership under President Xi Jinping acknowledged the pressing need for reform (the president quoted Deng Xiaoping's remark of 1992 that, without reform, the country would reach a 'dead end') and asserted the importance of the increased use of markets in industry and finance and of bringing greater clarity to the fraught issue of rural property rights.[24]

The fact that China has thus far been unable to bring about the rebalancing its leadership acknowledges it requires reflects the deep structural basis of the existing growth model and the determination of the CCP to retain its monopoly of political power. Effective rebalancing requires a further shift in investment activity from the state to the private sector and the liberation of the consumer. Both represent immense political challenges: the first because of the close symbiotic relationship between the leadership of the CCP and the SOEs already alluded to and the second because liberating the consumer implies some surrendering of economic control by the state. Some observers of China are of the view that the suppression of consumption in the economy is not an incidental feature of the system but a integral part of a growth model which transfers wealth from the household to the corporate sector by imposing low interest rates (thereby requiring people to save more to provide for their security), maintaining an undervalued currency, and providing cheap credit to business. Accordingly, the problem cannot be addressed without throwing out the whole growth model.[25]

For these reasons, the Chinese leadership has thus far balked at instituting the changes that would be necessary to make rebalancing a reality. Faced with a slowing economy, and concerned as always at the potential this holds for promoting social unrest, the leadership has continued to resort to policies of fiscal and monetary stimulus which worked in the past, albeit that these are producing ever diminishing returns. The challenges of effecting the necessary transition have been aggravated by developments in the real estate sector. This has thrived on the back of easy credit in the years following the GFC and, as one of the few available sources of internal economic growth, the government has been reluctant to restrain it. As a result, property investment had by 2014 come to account directly for around 13 per cent of GDP (and possibly 25 per cent indirectly), roughly twice the US share at the height of its property bubble in 2007. And, as in the United States, the financial system of China – both official and unofficial variants – had become highly vulnerable to a cooling of the property market. It has been estimated that property is used as collateral in at least two-thirds of total lending, while booming revenues from land sales have fuelled much local and provincial government spending.[26]

How these policy dilemmas resolve themselves is beyond the scope of this book. Notwithstanding the political challenges they represent, there are many who believe that the Chinese leadership, which has, after all, been remarkably skilful in its economic management over the past 20 years, will be able to bring about a controlled and gradual transition. But the possibility clearly exists that the transition, when it comes, will come in the form of some sort of crisis. Either way, the rate of China's growth in demand for mineral commodities is set to slow, this as much because of changes in the structure of China's economic growth and increases in the efficiency of mineral use as because of changes in the rate of economic growth itself. The key question for the mining industry is not so much whether growth in mineral demand from China will slow but how far and how fast it will slow.

Many analysts of the industry point to the fact that, for all China's recent achievements, GDP per head of population is still low in world terms and that the use of commodities per head of population still offers significant upside for demand over the long term. This may be so. But how it gets to the long term is going to be critically important. As this chapter has demonstrated, China has changed everything for the mining industry since the start of the millennium. What happens in China is pivotal to what happens next. Returning to the issue with which this chapter began, there remains a widespread conviction among iron ore producers that China's steel production will grow to over one billion tonnes a year before reaching a plateau.[27] It is just possible that they are making precisely the opposite mistake to that made by corporate analysts back in 2000.

3
Feeding the Dragon

Keeping China supplied

In May 2011, Brazilian iron ore producer, Vale, took delivery of the 'Vale Brazil', the world's biggest dry bulk carrier. Built by Daewoo in South Korea, the ship has a carrying capacity of 400,000 tonnes. At 360 metres long, it is 60 metres longer than the Eiffel tower is tall. Its carrying capacity is equivalent to the weight of almost 50 Eiffel towers. Along with the 18 other similar carriers on order, it was intended to bring about a step change in the cost of shipping Vale's iron ore from Brazil to Asia. Dry bulk carriers, as the name implies, are distinct from those designed to carry bulk liquids, such as oil tankers, and are used for the delivery of cargoes such as coal, iron ore, cement and grains. Up to this time, the largest vessels carrying these products were Capesize vessels, so called because their size prevents them from travelling through the Suez Canal, obliging them to sail around the Capes. Capesize vessels typically carry around 175,000 tonnes of cargo, only half what the new ships can carry. Their huge size means that there are only a few ports around the world which can accommodate a very large ore carrier (VLOC) as these ships are called. Among those that can are Dalian and Qingdao in China.

China's rapid and material-intensive growth brought with it a corresponding requirement for a rapid increase in the supply of mineral raw materials and an improvement in supply logistics. China's factories, power stations and construction companies were sucking up natural resources at an unprecedented rate. For producers of mineral products, who had experienced only modest growth in their markets over the preceding 20 odd years, this represented a major challenge, as well as an enticing opportunity.

China is a large country with significant mineral wealth. However, it was not well placed to meet the rising demand for minerals from its own resources in the time frame required. Although, as noted earlier, China was historically a major exporter of minor metals and industrial minerals, it does not have large, good-quality resources of some of the major metals on which industrialisation depends. It has limited, and generally indifferent, resources of bauxite (the ore for aluminium), copper and nickel. Its iron ore, while relatively abundant, is generally low grade. For some metals, such as platinum and cobalt, China does not appear to have any meaningful resources at all.

There is also a geographic dimension to the domestic supply challenge. Even where China has mineral resources, they are not always in the right place. Much of China's most vigorous development has taken place down its east coast, where it is commonly easier to bring in bulk supplies by sea rather than drag them overland across China's mountainous interior. China's coal production is concentrated in the north and northeast of the country while a large part of the need for this coal is at power stations in the south.

China's approach to this predicament has been pragmatic. With its low-cost labour and plentiful supply of capital, it generally sees its international comparative advantage as a manufacturer. As regards the supply of raw materials to its manufacturing industries, it has for the most part been happy for them to pick these up wherever they could be most competitively sourced. Price and availability of minerals is more important for sustaining China's competitiveness as a manufacturing nation than is their provenance. Certainly, China has not committed itself to the goals of self-sufficiency which marked the communist thinking of Mao Zedong or the Soviet Union in an earlier era.

However, if the state has been pragmatic with respect to the country's supply of raw materials, it has not been prepared to leave these matters entirely to the market. Mineral supplies are not just a matter of commercial importance for individual enterprises. Their availability plays a part in China's broader development ambitions and in the ability of the economy to grow and to create jobs. As such, it also plays a part in sustaining the position of the political elite. Minerals' availability is thus a strategic matter, a matter of political importance and a matter pertaining to China's economic and military security. It is axiomatic that shortages of minerals should not be allowed to compromise these strategic objectives.

Accordingly, the government has adopted policies designed to ensure that the necessary supplies are forthcoming, building infrastructure to facilitate the import and transport of raw materials, sometimes subsidising energy supplies for local producers of raw materials, and turning a blind eye to the environmental impacts of domestic raw material production. It maintains a strategic stockpile of important metals under the Strategic Reserve Bureau (SRB). Also, just as Japan and South Korea did before it, it has encouraged the local smelting and refining of metals even where the raw materials for this production have to be imported. To achieve this, it has sometimes sought to protect its metallurgical industries through the use of import tariffs and power subsidies. Finally, it has been anxious to see a diversity of mineral supplies. As the world's fastest-growing market for minerals, it does not wish to become too dependent on any single source of supply and does not want to see undue concentration of supply force up the prices its industries are required to pay for their raw materials.

Supplies for China's industries come from several sources, imports from foreign suppliers, domestic production and China's own overseas mining investments. The remainder of the chapter looks at each of these sources in turn, taking in a digression on China's trade policies along the way.

Imported raw materials

As China's industrialisation gathered momentum at the start of the 2000s, it increasingly had to turn to world markets to make up for the shortfall in its domestic raw material supplies. Markets were at the time generally oversupplied as a consequence of weak demand elsewhere in the world, and prices were at attractively low levels. The evolution of China's imports of some major mineral raw materials is shown in Table 3.1.

Iron ore

In line with its steel production, China's imports of iron ore had been gradually increasing through the 1990s. By 2000, its imports stood at 70 million tonnes. Although a substantial quantity, this still equated to only 15 per cent of the global market for seaborne iron ore, the same as the country's share of global steel production. Imports of iron ore into Japan at this time were higher, at 125 million tonnes, as were those into the European Union, at 110 million tonnes.

Over the course of the next decade, China totally transformed the character and prospects of the global seaborne iron ore market. (Seaborne iron ore is so described to distinguish it from iron ore traded

Table 3.1 China's imports of selected mineral raw materials, 2000–2013

	Units	2000	2004	2008	2013
Bauxite	Mt	0.4	0.9	25.9	71.6
Alumina	Mt	1.9	5.9	4.6	3.8
Copper concentrates	Mt	1.8	2.9	6.1	10.1
Refined copper	Mt	0.7	1.2	1.5	3.2
Copper scrap	Mt	2.5	4.0	5.6	4.4
Iron ore	Mt	70	208	444	820
Nickel ore	Mt	...	...	12.3	71.2
Refined nickel	kt	12	65	117	168
Thermal coal*	Mt	(–)	(–)	(–)	196
Coking coal*	Mt	(–)	(–)	3	76

Note: ... signifies negligible. (–) signifies exporter. * signifies net imports.
Source: China Customs, WBMS.

using overland transportation. It constitutes by far the largest element in globally traded iron ore.) As noted in Chapter 1, global consumption of iron ore had scarcely grown over the period 1970–2000. The market for seaborne iron ore, being less influenced by developments in the United States and Russia (which are largely self-sufficient in iron ore), had done rather better, but had still only grown at around 2 per cent a year over the period, hardly dynamic. Accordingly, this was considered a fairly unattractive business to be in. Growth was glacial, the real price of the product was on a declining trend and producers had to permanently battle their costs to maintain their margins

A few years on, things had changed dramatically. China's burgeoning steel production was sucking up iron ore from all around the world. As a result of the Zhu Rongji reforms discussed in Chapter 2, China's steel production, far from coming from just a handful of large integrated coastal mills, was beginning to flow in increasing quantities from numerous small privately owned mills dotted all around the country. This was not at all what the world's major mining companies had expected or prepared for. But it was having a dramatic effect on the demand for imported iron ore. By 2004, China's imports had grown to 208 million tonnes, by 2008 to 444 million tonnes and by 2013 to 820 million tonnes (Figure 3.1). It had massively outstripped Japan and the EU to become the largest importer of iron ore in the world, accounting for 65 per cent of global trade in the product.

The effects on the supplying industry were similarly dramatic. The principal suppliers to the seaborne market, Australia and Brazil, were

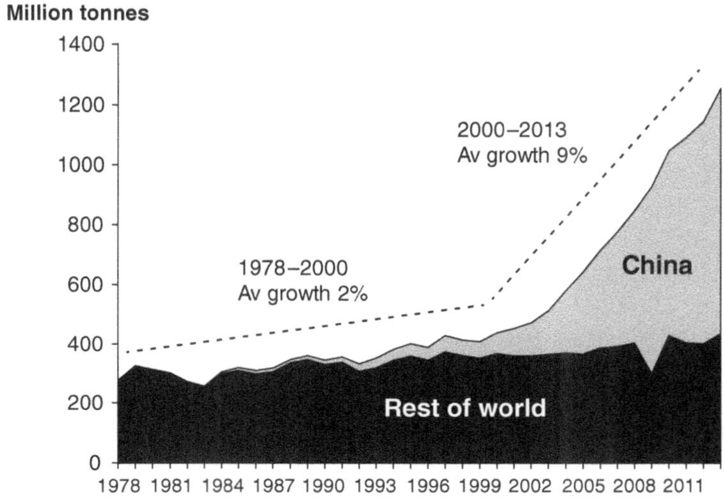

Figure 3.1 World imports of seaborne iron ore, 1978–2013
Source: China Customs and industry sources.

quick to ramp up their production to full capacity to meet the upsurge in demand. But the spare capacity was quickly exhausted and producers were soon pushing up against the buffers. Increasing production capacity at iron ore operations is a slow and costly process. Iron ore mining is essentially a materials-handling business, the barriers to expansion lying less in the acquisition of more trucks and shovels than in extending rail, pipeline and port infrastructure. This cannot be accomplished quickly. As producers approached the limits of their available capacity, iron ore prices began to surge.

With its major suppliers struggling to keep up with the rising demand for iron ore, China was obliged to look to other sources of supply. This supply mostly came from mines which were relatively small scale and did not require a lot of capital spending on them, being generally labour intensive. Although they had higher costs than the traditional producers, these newcomers could nonetheless make money at the higher prices prevailing at that time. One country that grew rapidly as an exporter of iron ore to China in these years was India, which by 2010 was supplying around 100 million tonnes, or 16 per cent of China's ore imports. Although concerns in India over the depletion of its resource base and the environmental impacts of its iron ore industry have subsequently seen India's share of China's ore imports fall sharply (exports in 2013 were 14 million tonnes), supply from a wide range of other countries, including South Africa, Iran, Ukraine, Russia, Chile and Indonesia, continued to grow and, by 2013, accounted for around 30 per cent of China's imports of iron ore. Australia, meanwhile, had a 48 per cent share and Brazil 20 per cent.

Copper

Even more so than in the case of iron ore, China is dependent on imports for its supplies of copper. As of 2000, copper from China's mines accounted for only around a third of its annual consumption of refined metal. The rest came from scrap or was imported. The explosion of demand which occurred in the early part of the 2000s, and the relocation of producers of semi-manufactured copper products to China to be near their customers, saw China transformed into a major buyer in world markets of all forms of copper. Given its limited capacity for smelting and refining copper at that time, the initial burden fell on supplies of ready-to-use, refined copper. Imports of such copper rose gradually from around 650,000 tonnes in 2000 to over 1.2 million tonnes in 2008, before jumping to over three million tonnes in 2009 as China seized the opportunity of abundant supply and weak prices to add to its commercial and strategic stocks of the metal.

China's preference, however, was generally to import the most basic mineral raw materials and process them domestically rather than buy the finished, refined product. Low capital costs and low labour costs meant that China could process these raw materials through to metal relatively cheaply, but government also deemed that there were strategic benefits to be had from the development of the country's metallurgical industries and supported these efforts through subsidised power and through tariff protections. (China had a 2 per cent import tariff on refined copper until 2008.) The importing of copper concentrates was not in the first instance so easy, since much of the world's supply of concentrates was tied up under contract. So, China had begun instead to import increasing amounts of copper scrap for re-smelting and re-refining. These imports rose from 2.5 million tonnes in 2005 to 4.5 million tonnes in 2005 before levelling out in the range 4.0–5.6 million tonnes a year.

Copper concentrates are the most basic tradable form of copper, typically containing around 30 per cent copper metal. As China built up its capacity to smelt and refine such materials, their flow into China increased. Standing at 1.8 million tonnes in 2000, imports of concentrates rose to ten million tonnes in 2013, turning China into comfortably the world's largest buyer. The principal sources of concentrates were

Chile, Peru, Australia, Mexico, Canada and the United States. Given the limitations of supply of these materials, China's buying had a material impact on the price smelters and refineries were obliged to pay for processing the raw materials supplied to them – so-called treatment and refining charges (TCRCs). When miners sell concentrates to smelters and refineries, the buyers negotiate a discount on the value of the copper contained in the concentrate to reflect the cost to them of converting it into finished refined metal. Since this is a negotiated price, it is subject to year-to-year variation, but over the period 1980–2005 the discount averaged around 23 per cent of the copper price. Under pressure of Chinese buying, the discount plunged to 5 per cent in 2007, making copper smelting and refining an unprofitable business for most market participants. It remained at this level until 2013, when finally new mine capacity started to come on and the market for concentrates began to ease.

Nickel

In the earlier stages of its industrialisation, China had relatively little requirement for nickel, which tends to be used in high-end products. Its focus was on more basic industrial commodities such as steel and cement. Even as late as 2000, a time when China was already consuming 17 per cent of world steel, it was only consuming 5 per cent of its nickel. Thus, while China had only limited domestic resources of nickel, its requirement for imports was minimal. In 2000, imports of refined nickel were just 11,600 tonnes.

However, as China began to focus its production more on value-added and lifestyle products, its demand for nickel began to grow. Through the decade, its imports of refined nickel increased steadily until by 2008 they stood at 116,900 tonnes. The same pattern was observable with its imports of ferronickel, an alloy of iron and nickel used directly in the production of stainless steel, which rose from virtually nothing in 2000 to 64,400 tonnes in 2008.

This rate of import growth for nickel products was not, however, enough to satisfy China's growing nickel requirement. As a result of years of poor prices and low levels of investment, nickel was a supply constrained market, and, under pressure of Chinese buying, nickel prices were being driven to spectacular heights. From an average level of $5945/tonne ($2.70/lb) in 2001, nickel prices rose to $13,845/tonne ($6.28/lb) in 2004, before exploding to $37,216/tonne ($16.88/lb) in 2007, a more than sixfold increase over the 2001 level.

China's response to this was to seek to add to supply by producing nickel from unconventional sources. It has long been known that it is possible to produce a low-grade ferronickel product from the direct smelting of pretty much any kind of nickel-containing oxide ore. However, the early application of the technology was dirty and inefficient and, as the understanding of metallurgical processes advanced, these earlier technologies were abandoned in favour of more sophisticated smelting technologies and the more selective mining of higher-grade nickel ores with low impurities.

Thus, it was that Chinese producers having access to redundant iron ore blast furnaces began to experiment with trying to produce a ferronickel using the old, abandoned, technology. In doing so, they found that they could produce a nickel-containing product, dubbed 'nickel pig iron' (NPI), which, although low grade and high in impurities, could nonetheless fulfil the requirements of certain less demanding applications for nickel, notably those in lower-grade stainless steels. Raw materials for this production could be readily sourced from Indonesia, the Philippines and New Caledonia where a lot of raw materials were already sitting on the surface in waste dumps as a result of earlier, more selective, mining operations. Although a relatively high-cost process, given the requirement to ship low-grade materials long distances and the process's requirement for high-quality coking coal, capital costs were very low. With the nickel prices then prevailing, producers could make money.

Over time, the sophistication of producing NPI rose, with producers shifting first to the use of electric arc furnaces and later to rotary kiln electric furnaces (RKEFs), and the costs of production declined from over $20,000 a tonne to nearer $12,000 a tonne. In the process, China became the world's largest producer of nickel. In 2013, its total output of nickel was 694,000 tonnes, of which NPI accounted for 482,000 tonnes, giving China over a third of global nickel production. To support this production, China's imports of nickel ores grew steeply, rising from effectively nothing in 2005 to 71 million tonnes in 2013, before falling away in 2014 as the effects of Indonesia's ban on unprocessed mineral products began to bite – an issue to be discussed in Chapter 7.

Aluminium

As with steel, China established itself during the first decade of the new millennium as the world's largest producer of aluminium metal. This confounded many in the aluminium industry outside China, who had long contended that China had no competitive advantages in the production of aluminium, having neither good local resources of raw materials nor cheap power (the two largest cost elements in aluminium production). Nonetheless, by virtue of the low cost of building plant in China and the availability of subsidised or stranded power (i.e., power, particularly hydroelectricity, which does not have any obvious alternative uses in the areas where it is generated) had not only enabled Chinese aluminium to be competitive but had resulted in a simply massive expansion of the industry so that by 2013 China was producing 22 million tonnes of aluminium a year, equating to 46 per cent of global output. This was 18 million tonnes more than the next largest producer, Russia.

To meet the rapid growth in its aluminium production, China was heavily dependent on imported raw materials, which is to say alumina, the white aluminium oxide powder used in the smelting of aluminium, and bauxite, the ore minerals used to produce alumina. While China has substantial resources of bauxite and is indeed a major exporter of bauxite used for producing heat-resistant bricks, this bauxite contains high levels of the mineral diaspore and is not generally considered suitable for the production of alumina.

As China's aluminium production expanded, increasing quantities of alumina were drawn into China from producers around the world. From imports of less than two million tonnes in 2000, these rose to seven million tonnes in 2005, at which time virtually half of China's alumina requirement was being supplied from abroad. From this point onwards, however, things changed, encouraged by the Chinese authorities who were becoming increasingly nervous about the country's high level of import dependence for this critical product. The first thing that happened was that some alumina refineries introduced an additional sintering stage in their processing to enable them to use domestic bauxite previously considered unsuitable for metallurgical purposes.

The other thing to occur was that China started to import more bauxite for processing into alumina. Over the years, less and less bauxite had been traded internationally since the economics of alumina production favour locating alumina refineries close to sources of bauxite supply. This is certainly the pattern in the world's largest and lowest cost producers of alumina such as Australia and Brazil. China's actions reversed this trend. In a move closely mirroring that for nickel ore, in 2005 China started to import increasing amounts of bauxite for processing in domestic processing plants. Starting out at two million tonnes in 2005, by 2013 these imports had risen to over 70 million tonnes. The largest supplier of this material was Indonesia, followed by Australia.

Through the use of unconventional technologies and by adopting policies which at the margin favoured the economics of the domestic processing of minerals over the importing of finished products, China was securing supplies on a scale and in a form that suited its purposes. In so doing, it was reshaping global patterns of trade in minerals and, in some cases, changing the form in which those products were traded. However, as will become apparent later on, it was also running directly counter to the aspirations of mining countries which wanted to do more of their own processing domestically. In particular, it was on a collision course with Indonesia, the principal source of its nickel ores and bauxite.

Other imported commodities

The import trends which can be observed for the metals discussed above were apparent across the whole range of minerals and metals that China required to support its economic and industrial development.

For a number of metals of which China is a substantial producer and of which it has in the past been a significant exporter, rapidly growing domestic demand first started to curtail exports then turned China into a net importer. Thus, China was up to the middle of the 2000s an exporter of lead, zinc and tin, but, by the end of the decade, it had to import these metals to meet its requirements.

Even for coal, for which China ranks as the world largest producer and for which it has historically been self-sufficient or a small exporter, it was increasingly turning to global markets to obtain additional supply. By 2009, China had become a significant net importer of both thermal coal (used in power stations) and coking coal (used for metal production). In the same year, it ceased exports of metallurgical coke (produced from coking coal), a product of which it had up to that time been the world's leading exporter.

China has negligible production of the platinum group metals; so effectively all the platinum and palladium it requires to supply its jewellery industry and its producers of auto catalysts have to be imported. China's imports of platinum and palladium account, respectively, for around 30 per cent and 20 per cent of global demand for these metals. China does have domestic resources of gold and is in fact the world's largest mine producer, but such is its demand for the metal that it is also a major importer.

The same pattern is evident for non-metallic minerals. As the world's largest producer of ceramic tiles, China's imports wholly dominate world trade in zircon. It is also a major importer of mineral feedstock for the production of the white coater and filler, titanium dioxide. With its huge population and limited agricultural land, China needs large amounts of fertiliser in order to feed its people. While it is a large producer of phosphates, with over 40 per cent of global production, its resources of potash are more limited and domestic production has to be supplemented by imports.

And so the list goes. Right across the spectrum of mineral commodities, China's growth has drawn in minerals from all around the globe. How its buying evolves will be a key factor shaping the future development of the global mining industry.

Domestic mining

'Plugging a cigarette into his mouth, He Shouming runs a nicotine-stained fingernail down a list of registered deaths in Shangba, dubbed "cancer village" by the locals. The Communist Party official in this cluster of tiny hamlets of 3,300 people in northern Guagdong province, he concludes that almost half of the 11 deaths among his neighbours this year, and 14 of the 31 last year, were due to cancer. Mr He blames Dabaoshan, a nearby mineral mine owned by the Guangdong provincial government, and a host of smaller private mines for spewing toxic waste into the local rivers, raising lead levels to 44 times the permitted rates. Walking around the village, the water in the streams is indeed an alarming rust-red. A rice farmer complains of itchy legs from the paddies, and his wife needs a new kettle each month because the water corrodes metal. "Put a duck in this water and it would die in two days", declares Mr He.'[1]

While imports have become critical to supplying China with minerals, a major part of China's growing demand for minerals has been met by production from China's own mines. Official data show investment in the sector growing at an average of 24 per cent a year over 2004–2013.[2] Driven by local industrial demand, and by the increases in mineral prices to which this has given rise, China has become the largest mining country in the world.

Table 3.2 shows how China's mine production grew between 2000 and 2013. It also shows how China's share of global production and China's world ranking as a producer have increased over the period.

Always a large producer of coal, China has continued to grow its output so that it now accounts for almost half the world's total hard coal output. China is also the world's largest mine producer of lead, zinc, tin, manganese, gold, tungsten and molybdenum. It is the second largest mine producer of copper (after Chile) and third largest producer of iron

Table 3.2 China's mine production, 2000 and 2013

		2000			2013		
		Output	World %	Rank	Output	World %	Rank
Coal	Mt	1384	30	1	3561	46	1
Bauxite	Mt	7.9	6	5	44.0	16	3
Iron ore	Mt	112	13	3	269	14	3
Manganese	Mt	5.3	25	1	14	29	1
Copper	kt	593	5	6	1707	9	2
Lead	kt	660	22	2	3048	54	1
Nickel	kt	50	4	7	98	4	7
Tin	kt	97	40	1	149	46	1
Zinc	kt	1780	20	1	5392	39	1
Molybdenum	kt	27	20	1	111	42	1
Tungsten	kt	30	83	1	85	88	1
Gold	t	175	7	4	438	14	1
Silver	t	1600	9	5	3670	14	3

Note: Data refer to metal contained in ore, save for coal, bauxite, iron ore and manganese ore, which are reported in gross weight of ore. Iron ore data have been adjusted in line with UNCTAD practice to make China's ore comparable with ore mined elsewhere in the world. Chinese iron ore is very low grade and is officially reported only on a 'run of mine' basis.
Source: EIA, GFMS, RMD, Silver Institute, UNCTAD, WBMS, WCA.

ore, bauxite and silver. This large production notwithstanding, it has not been enough to prevent China from having to import large amounts of mineral products, as discussed in the previous section.

China's mining industry is different from that in most other parts of the world. For one thing, the major producers of many minerals are state owned, or owned by local authorities.[3] For another, the structure of the industry is different. While many of the world's miners have been working towards a model of mining that emphasises economies of scale, capital intensity, the application of advanced technologies and high standards of safety, China's model looks like a throwback to earlier times. There are examples of modern, large-scale mining practices in China (e.g., China is an acknowledged world leader in longwall mining for coal), but the industry is generally characterised by large numbers of small labour-intensive operations. It includes a long tail of artisanal miners, part-time 'pick and shovel' operators often having other occupations such as farming, who dip in and out of mining as and when they can find something to mine and prices warrant them doing so. The largest iron ore mine in China, the Qidashan mine of Anshan Iron & Steel Group (Angang), is reported to produce only 3.5 million tonnes a year of iron concentrates.[4] Mines in Australia's Pilbara typically produce upwards of 20 million tonnes a year.

Exactly how many mines exist in China, it is impossible to say. The National Bureau of Statistics, in its Yearbook for 2013, records 17,000 mining enterprises (7869 of these coal mines) employing 6.31 million, but the cut-off for inclusion in the relevant table are enterprises with annual revenues of five million RMB ($790,000). Many mines fall below this level.[5] Also, and by the government's own admission, many mines are illegal and thus fall outside the statistical reporting system. One source reports 30,000 coal mines in China, of which 24,000 are small mines, and quotes the Ministry of Land and Resources stating that there are more than 60,000 mines in China which do not have licences.[6] The World Gold Council (WGC) reports that there are over 600 gold mines in China.[7] Quite, possibly, China has more mines than the rest of the world put together.

The underlying economics of these mines is distinctive. Often the ores being worked are inferior to those mined elsewhere in the world, while the scale of operations is too small to make them of interest to those with capital-intensive production methods. However, for low-tech producers having access to large amounts of low-cost labour, the economics of these mines can be made to work, at least while prices are reasonably high, taxes are low or non-existent and environmental costs can be ignored. The delivered cost of iron ore from one of the large-scale, highly mechanised mining operations in Australia is of the order of $40 a tonne. In China, many mines, accounting collectively for some 250 million tonnes a year of production capacity, could (as of late 2014) produce at costs between $50 a tonne and $100 a tonne. However, more marginal mines, accounting for a further 150 million tonnes of potential iron ore supply, have operating costs over $100 a tonne.[8] While iron ore prices exceed $100 a tonne, which they did for much of the period 2009–2013, then there was a place for some of these higher-cost producers in the market. Indeed, it could be argued that they had a positive influence on the balance of the market by acting as 'swing' suppliers – opening up when prices were high and closing down when they fell – something which large, capital-intensive operations cannot do.

Consistent with the fragmented and unregulated nature of much of China's mining, environmental and safety standards are low. The point is starkly illustrated by the quotation at the start of this section relating to the Dabaoshan iron ore and copper mine in Guangdong and to the cluster of smaller mines around it. This is far from an isolated case. As the quotation also suggests, provincial governments, in their quest to boost growth and create local employment, have often been complicit in allowing low standards to persist. Although a nationwide issue, the problems are particularly acute in the traditional coal mining areas in the north and northeast of the country, notably in the provinces of Shanxi and Shaanxi. These areas have extensive problems of groundwater and river pollution, subsidence and high levels of airborne particulate matter. Unsurprisingly, they have been the source of major health problems and social unrest. The coal industry is also responsible for the majority of deaths recorded by the mining industry caused, for example, by explosions, roof collapses and flooding. Official statistics show fatalities falling from over 6000 a year in the early 2000s to under 3000 by the end of the decade, but the vast majority of coal mines in China are small mines where reporting is extremely patchy.[9] What is not in dispute is that the numbers are high and far outstrip those of any other country.

In view of the economic model operated by China's mining sector and its record on the environment and safety, interest from foreign companies in investing in China's mining sector has not been very high, certainly nothing like it has been in downstream manufacturing sectors. The enormous difficulties that foreigners face in navigating China's labyrinthine administrative processes and in competing with local companies are further reasons discouraging inward investment in the sector. It is true that in the early part of the 2000s efforts were made by the Chinese government to attract foreign investors, perhaps in the hope that they would help raise standards in the industry. The World Bank launched a series of annual conferences in 1999 to catalyse this process. But these efforts were effectively abandoned in 2007 and the production of several metals, including tin, tungsten, antimony and molybdenum explicitly placed off-limits to foreign investment for strategic reasons.[10] The main exception to this tendency is gold, where a number of producers, many of them listed in Canada and Australia, have established operations in the country, generally in partnership with local Chinese producers.

The rapid and uncontrolled growth of the country's mining sector represents a massive challenge for the Chinese government. While local demand for minerals was growing strongly, it was reluctant to interfere too strenuously in the industry for fear of losing much-needed supplies of raw materials. Moreover, it could expect little co-operation from local authorities in its efforts to bring greater order to the industry since many of these authorities were themselves promoting mining and benefiting from the revenues it generated. While there have, over the years, been numerous central government initiatives to crack down on illegal, inefficient, small-scale polluting mines, all too often these mines restart as soon as the inspectors leave.

The period following the global financial crisis (GFC) brought a new resolve to the Chinese government's attempts to increase regulation of the mining industry. With mineral markets moving into better balance, and a growing need for China to move away from its pursuit of growth at any cost, the justification for turning a blind eye to the consequences of the industry's headlong growth evaporated. Resource taxes were increased to ensure that miners meet the full costs of extraction. Numerous press reports attest to enforced closures in the sector and to tighter permitting regimes. One example of such reports, taken from 2012, is indicative of the scale of this undertaking.

> Authorities in North China's resource-rich Inner Mongolia autonomous region halted 467 illegal mining projects last year in a region-wide overhaul mainly aimed to ensure work safety and environmental friendliness of the mining sector, officials said Saturday. The regional land and resources bureau checked about 9,000 mining projects in the months-long overhaul, halting 467 illegal projects, ordering 887 mines to suspend operations and permanently shutting down 73 mines. Officials of the bureau said that the mines either were unlicensed, harassed the life of local residents, or failed to properly compensate for the use of grassland.[11]

There is, in addition, growing concern about what the uncontrolled growth of China's mining sector has done to its resource base. The kind of mining that has been taking place in China has often been suboptimal from a technical point of view, with low-tech operators 'picking the eyes out' of deposits and rendering them uneconomic for large-scale mechanised mining. This resource dilemma is reflected in the fact that China's share of global production in many cases far outstrips its share of global resources. For example, while China accounts for some 46 per cent of world coal production and 9 per cent of world copper mine production, it is has only 13 per cent of global reserves of coal and only 4 per cent those of copper. In some cases the position is even more extreme. China reportedly produces 64 per cent of the world's fluorspar but possesses only 10 per cent of its reserves.[12] There are concerns, in other words, that China is getting through its resources too quickly.

Such concerns have generated recognition that China's approach to resource development might be working against its longer-term interests. In 2011, China's environment minister, Zhou Shengxian, acknowledged that 'the depletion, deterioration and exhaustion of resources and the deterioration of the environment have become serious bottlenecks constraining economic and social development'.[13] Of equal importance is the fact that rising standards of living in China are making the population increasingly intolerant of the environmental destruction which industrialisation in general, and mining in particular, has wrought.

A key plank in China's strategy for cleaning up and bringing greater order to the mining and metals sector is industry consolidation, this to be accomplished through merger and acquisition among existing producers and through the closure of marginal players. A major initiative on industry consolidation was launched by the State Council in 2006,[14] and it has been the subject of many official statements and programmes since. The stimulus programme of late 2008 contained provisions for accelerating the closure of surplus capacity in key sectors, including metals, albeit that it also contained a contradictory – and probably politically more pressing – objective of protecting jobs.[15] Power shortages of the sort that afflicted the country in 2010 and 2011 have provided a further rationale for China to try to rationalise capacity in this energy-hungry sector.

The object of industry consolidation is to focus more of China's mining and metals production into the hands of a small group of large-scale, state-owned producers. The plan for iron ore, for example, is to concentrate the industry in six to eight companies centred on the country's largest producer Ansteel Mining.[16] This should make the industry more amenable to regulation than in the past. It is clearly also hoped that larger, better regulated, companies will make for greater operating efficiencies, better resource use, higher environmental performance and a safer workplace. In short, while it has a long way to go, China is gradually trying to move its model of mining closer to that operated by the advanced industrial nations.

Trade policy

The main burden of sourcing the raw materials required for China's industrialisation has naturally fallen on imports and on domestic mine production. However, trade policies have also been deployed as part of China's attempts to ensure a sufficiency of mineral supplies.

Prior to the explosion of growth in China's demand for minerals post-2000, China was, as has been remarked, more often thought of as an exporter of minerals than an importer. In common with many other emerging economies, China was keen to promote exports to fund imports and to build up reserves of foreign exchange. Minerals were among the products China could readily sell abroad, and exports received support from energy subsidies and tax rebates. Competition among Chinese producers in the export market was often fierce, resulting in acute downward pressure on prices. Over the years, this led to numerous charges among mineral producers elsewhere that China was engaged in unfair trade practices and effectively 'dumping' mineral products onto world markets (i.e., selling them abroad at prices below those at which they were sold domestically). Among the mineral commodities that were subject to anti-dumping actions against China in the United States and the EU were crude mineral raw materials such as tungsten concentrates, fluorspar and magnesia and a whole raft of processed metals and alloys, including ferrosilicon, ferromolybdenum, silicon carbide, tungsten carbide, silicon metal, magnesium metal, manganese metal and various steel products.

During the course of the 2000s, the circumstances of China changed. As its domestic demand for mineral commodities began to accelerate, the logic of encouraging the export of mineral raw materials began to fade. As the world's fastest-growing industrial nation, China wished to retain more of its domestic production for use in its own domestic industries. There was a further reason for questioning the rationale for exporting. As China's living standards were rising, public and government tolerance towards the chaotic and polluting behaviour of the mining industry was diminishing. The squaring of this troublesome circle – ensuring supplies to its industry while asserting greater control over its own production – pointed to a policy of China restricting, rather than promoting, exports of mineral commodities. Such a policy might also, it was hoped, serve to allay concerns in Western countries over Chinese dumping.

Accordingly, as the decade of the 2000s progressed, China's trade policies began to shift. In November 2006, at the same time the government began pushing for the consolidation of the domestic mining industry, it sought to slow down the rate of mineral exports by imposing export duties on a long list of commodities. Ferroalloys such as ferromanganese and ferrochrome became subject to a 10 per cent export tax, while a tax of 5 per cent was imposed on a number of refined metals, including copper, nickel and aluminium. Coking coal and coke exports were also made subject to an export tax of 5 per cent.[17] For many of these commodities, the tariff rate was subsequently pushed higher, into the range

15–25 per cent. Obedient to these pricing signals, exports of mineral commodities from China slowed.

As commodity prices moved towards a peak in 2007 and 2008, and then surged again during 2010–2011, the policy challenge facing China deepened. As the only major industrial economy still growing in the wake of the GFC, the desire to retain domestically mined products within its borders became more pressing than ever. This applied to large-volume conventional commodities such as the base metals and ferroalloys discussed above, but it also applied to a range of smaller-volume, high-tech metals of which China had, over the years, become a major world supplier. Falling into this category were some rather exotic metals used in relatively small amounts in electronic products such as computers and mobile phones, and in green energy products such as wind turbines and hybrid electric vehicles. Among these metals were indium, gallium, germanium, beryllium and the so-called rare earth metals.

These last became a particular focus of public and policy interest. The rare earths are a collection of 17 metals with similar chemical properties which occur together in nature. By convention, the rare earths are divided according to their atomic weights into seven 'light' rare earth metals and ten 'heavy' rare earth metals. The two principal light rare earths, lanthanum and cerium – which, confusingly, are not at all rare[18] – are used mainly in catalysts for cracking crude oil and in auto catalysts for reducing exhaust emissions. Lanthanum is also used in the production of advanced rechargeable batteries employed in electric cars. Two others light rare earths, neodymium and praseodymium, are used in the production of permanent magnets for high performance motors. Among the heavy rare earths – which do live up to their 'rare' designation – dysprosium is used in permanent magnets while terbium, europium and yttrium are used mainly as phosphors, substances that create variations in colour and brightness in electronic display screens and in lighting.[19]

During the course of the 1990s and 2000s, China had grown to become the largest producer of these metals, accounting for over 95 per cent of global supply. Production of the light rare earths is focused in the north of China, primarily Inner Mongolia, where it is recovered at relatively large mines. The heavy rare earths, by contrast, come from a large number of small, low-grade deposits in the south of China, in the provinces of Jiangxi, Guangdong and Fujian. The geological nature of these deposits is very different from those to the north (they are a type of clay), and production has historically been much more fragmented. Moreover, because of the ammonium sulphate used to leach the rare earths out of the clay, but also because some mines generate radioactive waste, it is very environmentally destructive.

Rare earths rather crystallised the challenge that China faced with respect to its mineral policies and became something of a flash point for China's trade relations with the rest of the world. Having introduced a system of quotas for rare earth exports and eliminated VAT rebates on such exports in 2004, China in 2006 reduced the export quotas and introduced quotas on mine production of rare earths. At the same time, it implemented a 10 per cent export tax on rare earths, as it did on a lot of other metals, as noted above. The following year, it introduced a limit on the processing of rare earths and raised the export tax to 15 per cent for some rare earths and 25 per cent for others. The export quota was also reduced further. In 2010, the quota was cut by 40 per cent. This produced howls of protest from China's overseas customers and was followed by a steep run-up in rare earth prices.

From China's perspective, the imposition of production and export restrictions on rare earths looked like a pragmatic solution to a practical problem. Moreover, the measures were accompanied by a series of other supporting initiatives designed to bring about consolidation in China's rare earths industry, to use its remaining rare earth resource more efficiently, to stamp out illegal mining and to raise environmental standards. A 'white paper' laying out an integrated strategy for the industry was issued by the State Council in June 2012.[20] However, other countries saw things differently. They saw it as a policy designed to give China an unfair competitive advantage in the production of rare earth-containing products. The United States, the EU and Japan, who not so many years before had been complaining that China's trade policies were swamping their markets with unwarrantedly cheap supplies of minerals and damaging their mine producers, were now complaining that China's procurement policies were damaging their consumers by withholding supplies from them. They also claimed that, in doing this, China was in contravention of its obligations under terms of its membership of the WTO, a matter discussed in more detail in Chapter 8.

'Going out'

In September 2007, China's Export-Import Bank (Eximbank) signed an agreement with the government of the Democratic Republic of the Congo (DR Congo) to provide $8.5 billion for infrastructure development and for the construction and refurbishment of mines. In return, China was granted rights over the development of ten million tonnes of copper resources and two million tonnes of cobalt. The vehicle for the mine development was to be a Congolese incorporated company, Socomines (Société Congolaise Minière), 32 per cent owned by the Congolese and 68 per cent by Chinese state-owned companies (China Railway Group and Sinohydro).[21]

Pressure from the International Monetary Fund (IMF), which was concerned that the deal would compromise DR Congo's ability to service its existing debts, forced a reduction in the value of the deal to $6 billion in 2009. It also turned out that resources committed to the project contained only 6.8 million tonnes of economically recoverable copper rather than the hoped-for ten million tonnes. The project nonetheless stands as a good example of the infrastructure-for-minerals projects under which China has sought to secure access to resources in mineral-rich countries while at the same time contributing towards their economic development. The scaled-down project is scheduled to achieve its first production in 2015.[22]

The final strand of China's policies to secure supplies for its domestic industries was investment by Chinese companies in mining overseas. Through this means, China hoped to attain greater direct control over its supplies of imported raw materials. It perhaps also hoped that its investment would contribute more generally to the worldwide availability of minerals and at the same time reduce the market power of established industry players.

Through the 1980s and 1990s, there were sporadic examples of Chinese companies investing in resource projects abroad. In the 1980s, Sinosteel acquired a 40 per cent interest in Hamersley Iron's Channar iron ore project in Western Australia. In 1992, Shougang acquired the Marcona iron ore mine in Peru in a privatisation of state assets. Then in 1998, China Non-ferrous Metal Mining (Group) Company (CNMC) acquired the Chambishi copper mines in Zambia. However, there was nothing systematic about this process.

In 2004, with China's growth beginning to put substantial pressure on global mineral supplies, the Chinese government determined that it would extend its 'going out' policy – a policy originating in 1999, which encouraged Chinese companies to invest overseas – to the extractive sector and throw the weight of the various government funding institutions behind it. That same year, the state company China Minmetals launched a bid for Noranda, the 59 per cent owner of the large Canadian miner Falconbridge. The bid triggered a storm of controversy in Canada. Accepting that the Canadian government would never sanction the deal, early the following year Minmetals withdrew.[23]

After these inauspicious beginnings, the 'going out' policy took on a more gradualist tone, with Chinese companies focusing on early-stage development projects rather than on the acquisition of operating assets. The initial targets for investment were, not surprisingly, projects for producing the major industrial commodities in which China was not so well endowed, principally iron ore, copper and nickel. (More recently, coal, uranium, gold and alumina have been added to the list.) Coming late to the game of international mining, China focused its attention on some of the more challenging mining regions where the competition from other investors was less intense and where government-to-government deals of the sort it liked could give it leverage. In many cases, and for the same reason, it was forced to target second-quality resources, many of the best resources being already locked up in existing producers.

The major exception to this focus on more difficult mining geographies was Australia. Australia's relative proximity to China, its role as a long-standing supplier of minerals to China and its welcoming investment climate made it a natural destination for Chinese overseas mining investment. Also, of course, Australia hosted vast resources of iron ore, a commodity which China was obliged to import in large quantities. By building up a presence in Australian iron ore, China hoped to achieve direct control over a substantial proportion of its imported iron ore. The China Iron and Steel Association (CISA) said in 2011 that it would like to see the share of iron ore imports from China-invested operations rising from under 10 per cent to 40–50 per cent over the next five to ten years,[24] an objective reiterated by the Ministry of Industry and Information Technology (MIIT) in 2014. Also, it was hoped that by developing substantial production of iron ore in Australia, it could reduce the stranglehold on exports maintained by the two largest producers, Rio Tinto and BHP Billiton.[25]

Chinese companies had, as already noted, earlier engaged in partnerships with the major players for the development of iron ore mines in Australia. At the start of the new century, two more such deals were completed, with Chinese steel company Baosteel entering into a joint venture with Hamersley Iron (part of Rio Tinto) in 2001 and a consortium led by Wuhan Iron and Steel Group (Wugang) participating in a joint venture with BHP Billiton at Wheelarra in 2004. But by the middle of the decade, the emphasis was shifting to doing projects independently of the major producers, generally by buying into projects owned by smaller Australian companies. The list of such projects was long but included the acquisition by CITIC Pacific of the Sino Iron project at

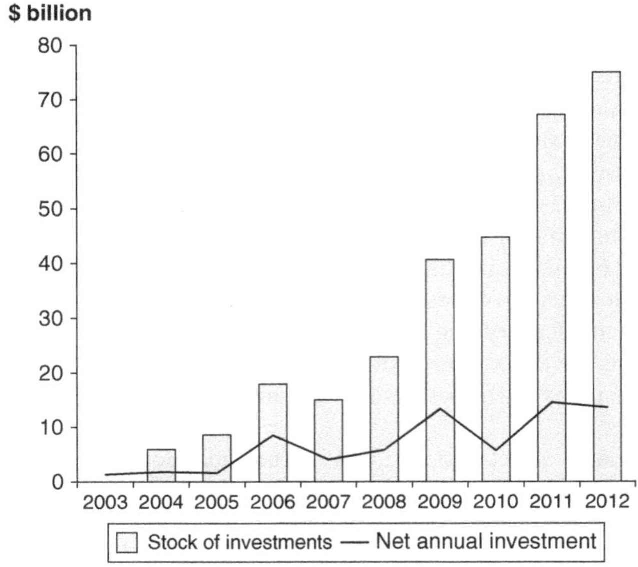

Figure 3.2 China's overseas direct investment in mining, 2003–2012
Source: NBS.

Cape Preston (2006), Angang's purchase of a stake in Gindalbie (2007), Shougang's (Shandong Iron and Steel Group) purchase of an interest (through Hong Kong-based APAC) in Mt Gibson (2008) and Sinosteel's purchase of stakes in Midwest Corp and Murchison Metals (2008).

Outside Australia, China has spread its net wide with its overseas mining investments. Official figures show foreign direct investment (FDI) in mining rising from $1.5 billion a year over 2003–2005 to an annual rate of $14.0 billion in 2011–2012 (Figure 3.2). They suggest that FDI in mining has in recent years accounted for somewhere in the region of 15 per cent of total Chinese FDI and that the sector accounts for a similar percentage of China's total overseas assets, making mining the biggest target of outward investment after oil and gas. Among the countries targeted for investment, neighbouring countries in Asia, including Mongolia, Kazakhstan, Pakistan and Afghanistan, have naturally featured. Latin America, with its large resources of copper, has also been a popular destination. Chinese companies are said to be behind a third of Peru's $60 billion pipeline of planned mining investments.[26] In South-east Asia, Chinese companies pursuing nickel have been drawn to Papua New Guinea (PNG), Indonesia and the Philippines. However, the region that has attracted most attention, not only from a mineral perspective but also politically, is Africa.

Africa is seen by China as a natural partner in its quest to develop additional sources of both oil and hard minerals. Africa, after all, possesses vast quantities of the very minerals that China most needs to import, that is, iron ore, nickel, copper and bauxite, as well as many others. Moreover, Western companies have generally been reluctant to commit to large new investments in Africa because of the lack of infrastructure or because of political instability. China, with its large state-backed companies and its ability to do government-to-government deals involving infrastructure-for-minerals of the sort described at the beginning of this section, seemed better adapted to do business there. As an emerging country itself, with a pragmatic attitude to doing business in other emerging countries, China evidently felt that it understood the challenges facing Africa better than many Westerners. Moreover, having engaged in a massive infrastructure build-out in their own country, China's construction and railways companies had precisely the sort of skills that Africa required for its infrastructure development. At the same time, Chinese development organisations and trade promotion bodies such as the China Development Bank and Eximbank were available to help bankroll the projects.

From the point of view of African host nations, China offered the prospect of capital and technology for development without the restrictive conditions and lecturing about governance which so often accompanied investments and aid from Western organisations. No less importantly, China had no colonial history with Africa. Indeed, its history was as a benefactor, having, for example, in the 1970s built the Tazara railway linking Zambia's copper mines to the coast through Tanzania. Accordingly, Africa became an important target for China's 'going out' policy. Politicians, accompanied by trade and industry missions, flitted between the two regions, affirming the importance of their relationship, the benefits which would stem from co-operation, and signing deals. Xi Jinping's first overseas trip as president in 2013 included three countries in Africa. The two-way flow of trade exploded from $10 billion in 2000 to $100 billion in 2006 and $200 billion in 2012. The African side of this trade comprised over 90 per cent oil and minerals. China's assets in Africa soared from to $10.6 billion in 2000 to $170 billion in 2013, with indirect holdings contributing a further $35 billion.[27] Along with the projects and the investment, large numbers of Chinese personnel arrived in Africa bringing skills and equipment with them. Across the whole continent, it was estimated that by 2013 there were around a million Chinese living in Africa.[28]

The copper of the African copper belt, in Zambia and DR Congo, was a natural target for China's investment. Chinese companies' involvement in the Chambishi mines and in the Socomines project has already been noted. In 2009, CNMC extended its interests in Zambia with the addition of the Luanshya mines and the Mulyashi project. More Chinese investment followed in 2011 with Jinchuan's acquisition of Metorex's copper–cobalt operations in Zambia and with Minmetals's acquisition of Canadian junior Anvil Mining and its Kinsevere copper project in DR Congo.

Iron ore in West Africa has been another target. In 2010, the Chinese state company Chinalco acquired a 45 per cent holding in the huge Simandou iron ore project being developed by London-based miner Rio Tinto. In addition to bringing much-needed financial muscle to the project, Rio Tinto, which had already been stripped of some of its iron ore concessions in Guinea for not developing them rapidly enough, clearly felt that the involvement of a Chinese company might offer it some political protections as well as provide assistance with the project's massive infrastructure requirement. At its Tonkolili iron ore mine in Sierra Leone, another London-listed company, African Minerals, partnered with three Chinese companies: a railway company, China Railway Materials (CRM), steel maker Shougang (which acquired 25 per cent of the company in 2011 along with offtake rights) and metal trader Tianjin Materials and Equipment Corp (which joined in 2013). In July 2013, it was reported that the government of Sierra Leone had struck a $8 billion infrastructure-for-minerals agreement with China covering a range of minerals including coal, iron ore, bauxite and rutile.[29]

Given the range of China's overseas interests and the speed at which they were accumulated, it is not surprising that some of these projects have had problems. As was illustrated by the case of Minmetals's attempted acquisition of Noranda, countries which are the target of Chinese investment are sometimes uneasy about the acquisition of resources within their territories by Chinese state companies. Several attempts by Chinese companies to buy Australian mining assets have had to be abandoned or modified after review by the country's Foreign Investment Review Board (FIRB) or following objections from shareholders. In Africa too, concerns have been raised about the continent's increasing dependence on China, with the governor of the Central Bank of Nigeria warning about 'a new form of imperialism'.[30] Michael Sata was elected president of Zambia in 2011 on an expressly anti-Chinese platform. In addition, it is clear that securing high-level government-to-government deals does not make Chinese-invested projects immune from local opposition. Mines in Zambia, Peru and PNG have been subject to direct action by local opponents of their mining projects.

Undoubtedly, many Chinese companies have found it harder than they expected to operate overseas. The fragmented nature of mining in China and the consequent lack of experience in operating large, highly mechanised mines cannot have helped. The approval process for development is often more complex and involves more stakeholders than Chinese companies are used to. Away from their home bases they lack their usual networks of known and reliable suppliers. Moreover, it takes time to adjust to different business cultures and attitudes to authority. Requirements to use local labour and to adapt to local labour laws means that they cannot always resort to the Chinese practice of throwing large amounts of cheap labour at problems as they are able to do at home.[31] The challenges are well illustrated by the Sino Iron project in Australia. This was characterised by misunderstandings over labour practices, large foreign exchange losses and disputes among its Chinese partners and eventually came into production $8 billion over-budget and four years late. The other three Australian iron ore projects mentioned above, Angang, Shougang and Sinosteel, have also wrestled with long delays and cost overruns.[32] As one observer put it, 'The challenge for China has been how to turn their talents as traders working off margin volumes to the different skill sets needed for foreign acquisition executions, business management and operating a business without government or political patronage.'[33]

While China's resolve to develop an overseas mining industry appears undiminished, the often bruising experiences of the past decade have led to some changes in approach. Rather than trying to buy its way into large mining projects over which it has total control, Chinese companies now appear more comfortable with the idea of taking minority stakes in projects in return for agreements to receive a share of the product. This was a strategy that worked well for Japan when it was trying to develop overseas interests in mining in the 1960s and 1970s. It is not necessary to own mines in order to benefit from their output. Chinese companies took advantage of the financial stress experienced by certain over-indebted large companies during the GFC to acquire substantial minority holdings. Thus, Chinalco acquired 9 per cent of Rio Tinto in 2008, while in 2009 China's sovereign wealth fund, the China Investment Corporation (CIC), acquired a 17.2 per cent holding in Canadian miner Teck and Hunan Valin acquired 16.5 per cent of Australian iron ore miner Fortescue.

There is also evidence of a more commercial orientation among Chinese companies operating overseas. While early investment emphasised securing raw materials to support China's domestic industrial development, the state authorities have clearly become frustrated by the poor performance of China's mining assets overseas (very few have been profitable), and companies are being required to look more closely at economic returns. Some have shown that they are prepared to walk away from projects rather than pay too much.[34] Consistent with this more market-sensitive orientation, China has been signalling that it will in future also encourage and support its private companies to go abroad, issuing a regulation on the matter in July 2012.[35] Hitherto, overseas investment in mining has been largely the preserve of state and state-backed companies.

The manner in which China has burst onto the global mining stage has generated concern in some quarters that the country is engaged in a resource grab, which will precipitate shortages and send commodity prices spiralling upwards.[36] However, the reality is that China's overseas mining interests, while they are certainly growing, are still small in global terms, and their impacts often exaggerated.[37] But China is in this for the long haul. In late 2011, it reaffirmed its commitment to the 'going global' policy and earmarked a further $580 billion over five years to support it.[38] On the basis of history, maybe $90 billion of this can be expected to go to mining. As has been emphasised, mineral procurement is a strategic matter for China and it clearly plans to become a power in global mining. Its huge domestic demand for minerals gives it a strong vested interest in seeing global markets well supplied. And it has enormous financial firepower, including that of its Sovereign Wealth Fund, the CIC, to put behind these efforts.

As China has discovered, there are no shortcuts to the process of developing offshore mines. Large, high-quality mining operations in most jurisdictions take a long time to develop, and China's record of project delivery thus far has been poor. State support for Chinese mining companies may be helpful in gaining access to resources overseas, but the development of long-term sustainable operations still requires that companies attend to the minutiae of the business: good operating practices, high environmental standards and respect for the workforce. In this, Chinese mining companies are no different from any other. As several Chinese companies have discovered to their cost, authoritarian management styles do not transplant well into other countries, and Chinese miners have to adapt to the characteristics and business cultures of their host countries if they want to succeed. Chinese companies, like any other, can be disrupted by the activities of disaffected local communities, and are vulnerable to the effects of resource nationalism. For China to build out its presence on the global mining stage, its companies will need to look like and behave more like those from other countries which have been in the business of global mining a whole lot longer.

4
Prices Take Wing

Market mechanics

The five-minute 'ring' starts slowly, with the dealers slouched on their red leather benches in a circle under the bright lights. Periodically, one of them proffers a trade with an air of studied indifference. This continues for four minutes. In the last minute of the ring, the pace begins to quicken. The bids and offers flow more freely. The air of indifference evaporates. Dealers become more animated. Outside the ring, telephones clamped to their ears, traders signal to their colleagues in the ring the buy and sell orders coming down the line from clients. The last seconds of the ring are a complete frenzy, the dealers straining forward on their benches shouting and gesticulating. Then, as quickly as it arose, the storm abates. The bell sounds and the shouting ceases. The dealers leave the ring and calm prevails. The price of metal for the next trading period has been set.

This is the strange and seemingly chaotic manner in which prices of some of the most important metals, including aluminium, copper, lead, nickel, tin and zinc, are set – or 'discovered' – on the London Metal Exchange (LME). In an age of telephones and computers, it all seems rather primitive. But the metal trade likes it and it works. There are telephonic and computer-based systems available to back up and complement the 'open-outcry' trading in the ring, but it is in the ring that the official prices of the LME are established. These are then used as the reference prices for metal and metal ores traded globally as well as for the settlement of LME contracts used by producers, consumers, traders and speculators.

Commodity markets are viewed by economists as among the most technically 'efficient' markets. That is to say, prices in the market are generally an accurate reflection of the relative pressures of underlying supply and demand. There are numerous participants in these markets, ensuring that they are highly competitive and liquid. They are also open and transparent. Most of the information bearing on supply and demand, such as trends in the world's major economies, disruptions to mine production or the size of metal stocks held in warehouses, is readily available to all market participants. In addition, commodity markets do not need to worry unduly about geography or branding. Marketing and distribution costs typically constitute a small proportion of the transaction price, while one lump of metal is pretty much indistinguishable from another of the same weight. All that most buyers or sellers in the market are interested in is the price. This makes these markets an economist's dream. You can draw supply and demand curves, and they work.

In addition to providing a means of price discovery for immediate, or 'spot', deliveries of metal, the LME and commodity markets like it offer the possibility of trading on future dates using forward contracts and options. This provides a basis for producers, consumers or traders to hedge their price risk by locking in their margins at the time of making a sale or purchase and an opportunity for profit for those who want to bet on the direction of metal prices. Since these forward contracts and options are paper contracts which are easily traded, the amount of metal traded annually through the market is many times the amount of metal traded physically. LME turnover is typically around 40 times annual metal production. The paper trade gives the market liquidity and thus helps the process of price formation and risk management.[1]

This chapter examines the behaviour of mineral prices in recent years and the role played by China in shaping this behaviour. It considers the idea that strong demand for commodities from emerging economies would result in a sustained commodity 'super cycle'. It goes on to look at the growth of commodity funds and to consider their impact on mineral prices before concluding with a discussion of how the rise of China changed pricing in the iron ore market.

Things begin to stir

As China's relentless and material-intensive growth, described over the two previous chapters, persisted, the effects began gradually to feed through into commodity prices. As discussed in Chapter 1, when the world entered the twenty-first century, prices were at painfully low levels, pressing down hard on producers' costs. But in the course of 2003, something began to stir. With the technology boom now burnt out – the NASDAQ had fallen to a quarter of its 2000 peak – the investment community was looking for the next 'big idea'. A prime candidate for this role was China and its impact on commodities. Through the economic trough induced by the burst of the tech bubble, China had continued to march on. Moreover, China's consumption of mineral commodities was now reaching a scale where it had a material impact on global markets. Adding to their conviction that markets were about to lift themselves off the bottom, analysts had noticed that the global demand for commodities appeared to be outstripping growth in industrial production, something that normally happens as manufacturers build stocks ahead of a cyclical recovery.[2]

By the second half of 2003, practical evidence of global economic recovery was beginning to emerge. Prices of most metals were moving upwards, some of them, notably nickel, quite strongly. They were also switching out of contango into 'backwardation' – a condition in which spot prices rise above futures prices – generally a sign of tightness in the physical market for metal. However, there was nothing in the markets at this stage to suggest that anything unusual was taking place. Clearly, the global economy was beginning to accelerate again after the cyclical trough of 2001–2002, and some sort of recovery in metal markets might have been expected. Prices were, however, well within the bounds of normal historical experience. For the most part, this continued to be the case in 2004. As global growth accelerated, metal markets moved into deficit and prices trended higher. In the gold market, producers were starting to abandon the practice of systematically selling forward their production and some were even buying back their hedged positions in anticipation of higher gold prices.

If China's rapid growth was continuing to fuel speculation about the possibility of a sustained rise in commodity prices, there were plenty of sceptics too. A rapidly growing trade deficit in the United States was giving rise to concerns over the sustainability of growth there, while in China the authorities were raising interest rates and clamping down on credit growth in response to fears that its economy was overheating. Reflecting these uncertainties, metal prices in 2004 displayed extreme volatility. With growth in the global economy projected to slow in 2005, and with the production of many commodities beginning to respond to the higher prices, some believed that 2004 would mark the peak of the cycle. The experience of the previous 25 years has taught managers in the mining and metals industry to be cautious.

Coming of the super cycle

Whatever misgivings there may have been about the prospects for the industry at the time, in retrospect it is clear that 2005 marked a major turning point for mineral markets. Early in that year, Alan Heap, a marine biologist turned metal analyst at the bank Citigroup, published a piece of research in which he argued that the mining industry was on the threshold of an upswing, not of a regular business cycle lasting maybe five–ten years but of a sustained, multi-decade 'super cycle'.[3]

The primary driver for this super cycle, Heap claimed, was 'higher trend growth in global demand, driven by materials intensive growth in an urbanising and industrialising China'. This higher trend growth would in turn entail 'new supply from higher cost projects, which drives industry average costs higher'. The combination of these factors, stronger demand and higher costs, would produce 'trend rises in real commodity prices, reversing the trend decline of the last 30 years'. Parallels were drawn with earlier periods of upward trending commodity prices, notably that which accompanied the urbanisation and industrialisation of the United States in the late 1800s and early 1900s and that driven by post-war reconstruction in Europe and Japan which led up to the commodity price spike of the 1970s. Heap believed that these developments pointed to the need for a major upward revision in the bank's long-term price projections for mineral commodities.

Heap's notion of the super cycle was essentially an empirical observation based on what was going on in the market at the time and on the evolution of prices historically. It also happened to fit rather well with the narrative about the growing importance of emerging markets and of their rising metal intensity and to be a useful idea on which to hang a marketing campaign for commodity stocks. The idea found support in academic circles. Econometric studies by John Cuddington and Daniel Jerrett found 'considerable evidence of ... past super cycles in real metal prices, defined here as cyclical components with expansion phases from 10 to 35 years'. They also found 'statistical evidence ... consistent with the claim ... that metal prices entered the early phase of a super cycle at the beginning of the 21st century'.[4] Not surprisingly, the theme was eagerly seized on by other analysts, investors, project promoters, mining ministers, company executive and the mainstream press.[5]

The months following the publication of Heap's research note added credibility to the notion of the super cycle. Despite 2005 being a lacklustre year for global demand, miners struggled to supply the market. Producers of copper, for example, collectively fell short of their targets for the year by one million tonnes. The price of copper moved steadily upwards to reach levels not seen before in nominal terms. Among the bulk commodities, iron ore and coal, commodities which had particular significance for China because of the strong emphasis of its growth on construction and power, there were some quite startling developments. For the financial year starting April 2005 (the conventional period for annual contracts in the Asian market), iron ore producers extracted a wholly unprecedented 71.5 per cent price increase from steel mills. The largest price increase prior to this was 19.5 per cent, in 1980. Producers of coking coal went one better with a price increase of 119 per cent.

What happened next, however, surprised even the most bullish observers of the market. In the last quarter of 2005 and first quarter of 2006, the prices of virtually all metals took a dramatic step up. Between September 2005 and May 2006, the average monthly price of copper more than doubled from $3857 a tonne to $8044 a tonne. At the same time, the backwardation in the market ballooned out, the differential between the spot copper price and the 27-month price in May 2006 exceeding $2000 a tonne. Gold went from $456 an ounce to $677 an ounce. It was as if all the arguments regarding China's growth and the constraints on supply had hit home at once. These prices were simply way off the scale of recent historical experience (Figure 4.1).

This situation persisted through to the middle of 2008. While prices fluctuated, they remained at historically elevated levels. Some managed to push even higher. Nickel's vertiginous monthly peak of $52,161 a tonne came in May 2007 while copper's peak (at least during this particular phase of the market's development) came in April 2008. Gold peaked at $971 an ounce in March 2008. Iron ore producers, after a couple of more modest price rises in 2006 and 2007, excelled themselves with a price increase of 79.9 per cent for the financial year commencing April 2008. Aluminium alone remained generally unmoved.

The collapse of Lehman Brothers in September 2008 and the accompanying global financial crisis (GFC) posed a serious challenge to the notion of the super cycle. Metal demand fell back sharply and prices went with it. The hiatus, however, proved relatively short lived. As described in Chapter 2, as a result of its economic stimulus programme, China's demand for commodities came storming back in 2009 and was supplemented by some opportunistic rebuilding of commercial and strategic stocks. With China by now the largest consumer of minerals in the world, this made a huge difference to commodity markets and largely offset the depressed condition of demand in the advanced economies. Certainly, it marked a huge difference with earlier cyclical

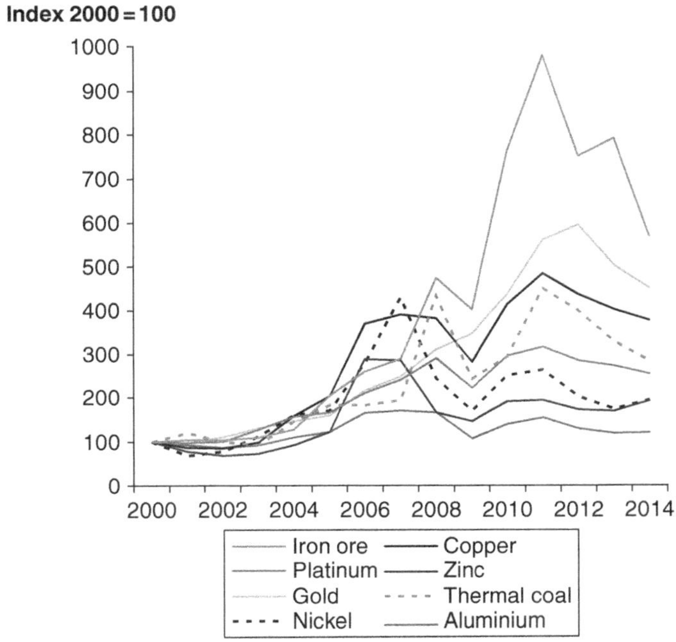

Figure 4.1 Prices of selected mineral commodities, 2000–2014
Source: LBMA, LME, LPPM, industry sources.

downswings such as those in the early 1980s and early 1990s, when the consumption of minerals fell sharply and stayed down. Equally to the point, the supply of many commodities was still constrained by limited production capacity. Although miners had been investing hard, it takes many years to bring on new mine capacity. A large mine can take eight to ten years to get up and running, and some take a lot longer still. In addition, as tends to happen in industries where existing plant is being pushed to the limit and new capacity is being added, there is a tendency for producers to fall short of their targets. Throughout this period, there was a constant gap of around one million tonnes between what copper miners were promising each year and what they were delivering.

Galvanised by China's spectacular recovery, many mineral commodities experienced a further leg-up in prices following the GFC, as is shown in Figure 4.1. Copper moved to a second and higher peak in early 2011, with a monthly average price in February not far short of $10,000 a tonne. Iron ore prices, which were knocked back in 2009, similarly surged higher to a new peak in 2011. The precious metals, gold, silver and platinum, notwithstanding their status as 'safe haven' investments, stalled along with the industrial metals in the 2008 crisis, but then began to rally in 2009 as central banks the world over stoked up inflationary expectations by feverishly printing money – so-called quantitative easing (QE). Gold and silver moved to new peaks in 2011 and 2012. Other metals, including aluminium, nickel, zinc, cobalt and molybdenum, rallied subsequent to the GFC but failed to reach their 2007–2008 price highs.

Although the experience varied according to commodity, from 2011 onwards prices for most commodities began to trend downwards and by late 2014 were back to where they had been five years earlier in the wake of the GFC. The world outside China was still struggling to regain momentum despite the aggressive use of monetary policy in the advanced economies. In China, the economy, although still growing, was growing at a slower rate. GDP growth, which was 10.3 per cent in 2010, fell to 7.4 per cent in 2014. In addition, the huge wave of investment, which had been going on in the mining industry since prices took off and which reached a peak in 2012, was finally beginning to come through in the form of new production. Markets were moving into better balance.

Was the super cycle real?

Much has been written on the super cycle and its nature. Did it represent a structural shift in the market or was it just an unusually long and strong cycle? Frequently the discussion on these matters is confused by the fact that the super cycle means different things to different people.

As noted above, Heap anchored his idea of the super cycle in two specific factors, rising demand from emerging economies, notably China, and constrained mineral supply.

Looking at the first of these two factors, rising demand, the evidence of an inflection in trend growth is somewhat mixed. It is clear that the surge in China's demand for materials used in construction, such as steel and aluminium, had a visible impact on growth in global demand for these commodities. The global iron ore market was clearly transformed by the acceleration in the use of steel in China.

However, as discussed in Chapter 2, the evidence from other commodities is less compelling. China's growing demand for copper and nickel resulted in only the most modest acceleration in global demand for these commodities. For the platinum group metals (PGMs), there was no visible impact on the trend in consumption. For these commodities, a significant part of China's growth appears to have come at the expense of demand growth elsewhere in the world as manufacturing industries relocated towards China.

What about the second part of Heap's explanation for the super cycle, constrained supply and rising industry costs as a driver of higher prices?

The extent to which rising demand for commodities impacts prices depends on the ability of the supply side of the industry to respond (the 'elasticity' of supply) and the cost at which it can do so. Large-scale producers of iron ore have a limited ability to respond to increases in demand once their spare capacity has been used up. The expansion of iron ore capacity commonly involves large infrastructure projects such as railways and ports, which have very long lead times. This goes some way towards explaining the extreme price response induced by China's burgeoning growth for iron ore. Aluminium provides a contrasting example. Although, in common with iron ore, it was one of the few commodities to show clear evidence of a trend shift in global demand resulting from China's growth, it did not experience the same price increases. The reason for this was quite simply that, alongside China's growth in demand for aluminium, China's production was also growing fast. Aluminium thus demonstrates that demand pressure from emerging economies was, in itself, an insufficient basis on which to mount an explanation for the super cycle.

For many commodities, it was certainly the case that the rise in commodity prices from the mid-2000s onwards was associated with a strong rise in costs.

A key tool for the analysis of the mining industry is the production cost curve, which serves as an approximation of an industry supply curve. It is created by plotting producers' output (or potential output) in order of average production costs, starting with the lowest cost producer first. The resulting line normally describes a gentle upward bending curve, which gets steadily steeper as the physical limits of the industry's capacity are approached and unit costs of production tend higher. At some point the curve goes vertical and, in the short term at least, no further supply can be made available. When the demand curve hits this steep, or vertical, part of the curve, the resulting high prices are a strong signal to producers to invest and to create more capacity.

As prices began to spiral upward in the mid-2000s, rising costs were evident everywhere in mining. Cost curves shifted northwards as the costs of inputs to mining increased. Shortages of skilled labour, equipment and parts resulted in higher wages and increased costs of supplies. Governments also used the opportunity of higher prices to extract additional rent by increasing royalties and other imposts. Rising commodity prices themselves contribute to rising costs of producing commodities by way of mine purchases of diesel, chemical reagents and steel. In addition, rising commodity prices boosted the exchange rates of commodity producing and exporting countries raising their costs in US dollar terms.

As well as shifting northwards, cost curves also steepened. In part, this was a product of the differential impacts of some of the factors mentioned above (some currencies are more sensitive to commodity price increases than others) but it was also the result of higher-cost producers being drawn into production as a consequence of higher prices and of existing high-cost producers, who might otherwise have expected to cease production, being incentivised by higher prices to stay in operation. The combined effects of these influences on the cash costs of producing copper between 2004 and 2007 are shown in Figure 4.2.

Rising pressure on costs was not confined to operating costs. The cost of creating new mine capacity was also increasing. A consequence of a large number of mining companies all trying to expand capacity at the same time was that the cost of most of the elements required to build

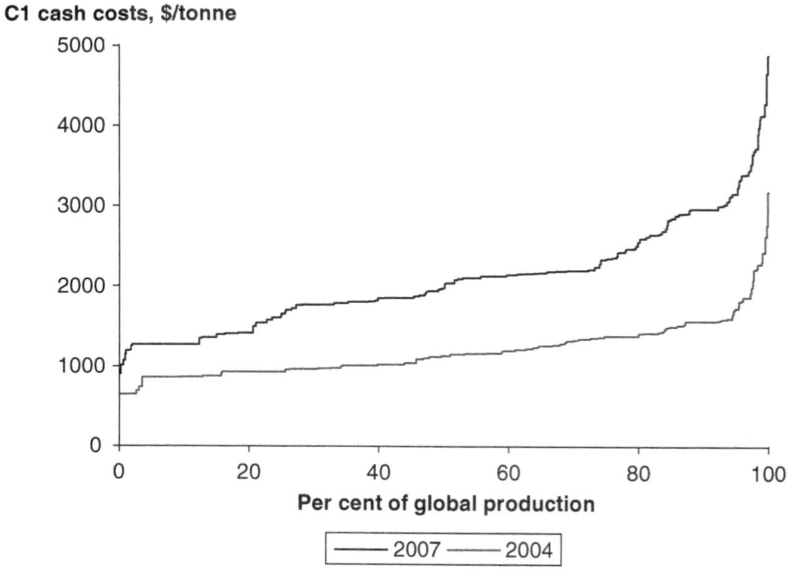

Figure 4.2 Copper cost curves, 2004 and 2007
Source: Wood Mackenzie (2008).

new mines, including project managers, skilled labour, engineering and design companies and heavy equipment, was being bid up. Moreover, in order to open up new resources, companies were being obliged to go to more remote or more marginal deposits where the costs of development were inherently higher. Here also, governments were adding to cost pressures with increased costs of permitting and environmental compliance. For years, a rule of thumb in the mining industry was that an annual tonne of copper mine capacity cost around $5000 to build. By 2010, many large projects had capital costs in the range $10,000–$15,000 a tonne of capacity. Some were pushing up through $20,000 a tonne. The capital cost of constructing a greenfield iron ore mine had similarly risen from around $100 a tonne of annual capacity to anything up to $200 a tonne and beyond. The rising costs of creating new capacity were not only adding to the long-run costs of producing mineral commodities but also adding to the uncertainties companies and their bankers faced when evaluating projects.

Unfortunately, none of these factors – whether on the demand or supply side – were in and of themselves a very satisfactory basis on which to mount the case for a 20-plus year price cycle. Surges in demand do drive up prices while costs, in the normal course of things, tend to ride up with prices. Generally, however, the market responds. Supply increases and prices and costs come back down again. Given the capital intensive nature of mining, the supply adjustment necessarily takes time. However, as of the late 2000s, there was nothing to say that this cycle was fundamentally different from those that had gone before. It may have been that the cycle was a particularly strong one and the supply lag therefore unusually long, but many informed observers of the industry were still of the view that the cycle was otherwise a conventional one and would work itself through within a decade or so, and that when it did so, prices would fall back towards historical levels.[6] To argue that this was a multi-decade phenomenon required demand to outrun supply on a more or less continuous basis over 20 years or more, keeping markets permanently under tension and high-cost producers in production. To these observers, this seemed improbable.

During 2012 and 2013, as prices of many commodities trended downwards, the view that the super cycle was really just a particularly strong and attenuated cycle attracted increasing support. In truth, some in the analyst community had never entirely bought into the super cycle theory, preferring the alternative, more pragmatic slogan, 'stronger-for-longer'. Gradually, the trickle of articles pronouncing the end of the super cycle became a torrent.[7] As prices slid lower and industry margins came under growing pressure, even Citi, Alan Heap's employer (Heap himself having tragically died in 2011), proclaimed the 'super cycle sunset'.[8]

An alternative view

If there was no super cycle in quite the form in which Heap envisaged it, the case can nonetheless be made that something important changed with the revival of commodity prices in the early 2000s. Prices may have fallen back from their peak levels, but they have not fallen back to anywhere near the levels they were running at prior to the commodities boom. It does appear, as Heap maintained, that the 30-year downward trend in real commodity prices was broken at this time. In this sense, it may well be that 2001–2002 marked a structural break in mineral prices.

One logical possibility for this state of affairs is that costs have gone up on a permanent, once for all, basis. The jury must be considered still out on this. Over time, it will no doubt become clearer what increases in costs during the past decade were cyclical and what (if any) were structural and therefore longer lasting. To argue that mining costs will rise on a sustainable basis runs counter to a lot of history. In the past, the productivity of the mining industry, while it has ebbed and flowed with the price cycle, has generally been able to offset the cost-increasing effects of mineral depletion. And certainly it does not do to underestimate the ingenuity of the mining industry. Research conducted by John Tilton provides evidence of a substantial decline in mining industry productivity (and increase in costs) during the years of the commodity boom, but, Tilton contends, this was largely, if not entirely, cyclical.[9] There is, however, no iron law which says that productivity gains must always outrun the effects of depletion. Although the situation inevitably varies from commodity to commodity, the case that depletion is pushing real mining costs higher on a sustainable basis may be stronger than technological optimists like Tilton allow. The subject is discussed, and the case is made, in Chapter 9. There also exists the possibility that the end of the era of cheap energy (or the start of an era of dearer carbon) could be a material factor in halting the downward trend in long-run mineral prices, given that energy accounts for around a quarter the cost of producing metals.

Aside from rising cost pressures, the other possibility is that the upward step in real-term prices which has occurred is rooted in factors wholly outside the mining sector.

Movements in the 'real' price of a commodity are simply a measure of how the price of that particular commodity has moved relative to the prices of commodities in some general economic variable such as GDP, consumption or production. Thus, a rise in the real price of copper could logically result from the nominal price of copper rising faster than the prices of the broader basket of commodities or it could result from the prices of those other commodities falling relative to (or rising less fast than) the price of copper. These 'other commodities' against which copper is being evaluated are typically dominated by manufactured goods and services.

It is now widely understood that the integration of China and India into the global economy from the 1990s onwards delivered an immense supply-side shock. In combination with other emerging countries, they added 1.5–2 million to the labour force engaged in international markets, more than doubling its size.[10] In addition, because of the high savings rates of many Asian nations, these labour-rich economies came with a lot of cheap capital. The combination of these factors massively suppressed the cost of producing manufactured goods. To a lesser extent (because they are not so transferable across national borders), it also impacted the cost of services such as call centres, the writing of software and back office services. The potential implications of this were spotted early on by Andy Xie, then Morgan Stanley's Asian economist, who argued in 2003 that 'China's size and development speed represent a discontinuity in the global economy. It permanently devalues labour against scarce global [natural] resources.'[11] At around the same time, Martin Wolf in the *Financial Times* suggested that the 'almost limitless supply of cheap labour' in China foreshadowed a major shift in the terms of trade between manufacturers and commodity producers in favour of the latter.[12]

An examination of terms of trade data for the decade after these statements were made show that this is precisely what occurred. Figure 4.3 shows IMF data for world prices of traded manufactures and commodities since 1990. They reveal that commodity prices, as might be expected, are more volatile than the prices of manufactures, but they also show that in the period following 2000 commodity prices rose very much faster than the prices of manufactures. The same shift is also evident in IMF data on terms of trade for developing Asia (predominantly an exporter of manufactures), which have deteriorated consistently since the mid-1990s. Conversely, they show that the terms of trade for Australia (predominantly a commodity exporter) have increased strongly since the late 1990s following an extended period of decline.

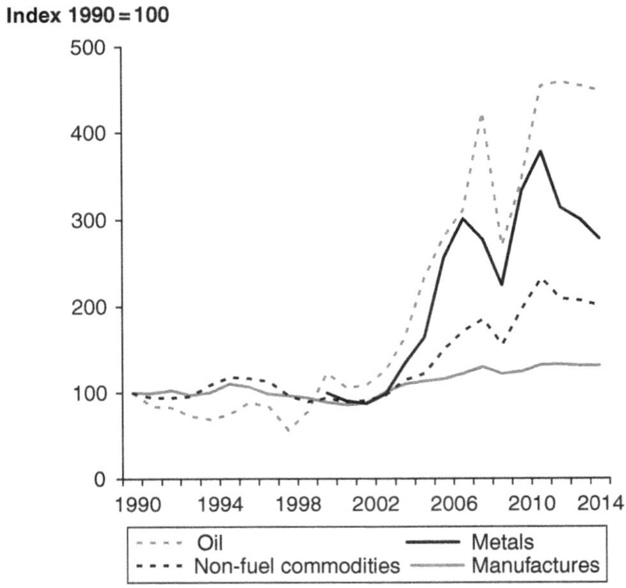

Figure 4.3 Prices of internationally traded products, 1990–2014
Note: IMF price data for metals start in 1999; so, the series shown here is based on 2000=100. Non-fuel commodities include food, beverages, agricultural raw materials and metals.
Source: IMF, World Economic Outlook, editions 1998, 2002, 2007, 2014.

In other words, the data suggest that the change in the terms of trade between manufactures and commodities was already afoot before the 2003–2008 commodity price boom kicked in and, therefore, likely had its origins in factors outside the commodities sector. Significantly, the quantitative analysis by Cuddington and Jerrett referred to earlier also points to the structural shift coming in the late 1990s rather than in the 2000s.[13]

A final factor, which has likely served to shape the trend in commodity prices over the past decade or more, is the unit in which mineral prices are commonly measured – the US dollar. Where the value of the currency in which a commodity is priced declines, the price of that commodity has a tendency to increase. During the 30 or so years that dollar prices of minerals trended down, the US economy was the dominant economy in the world and the dollar unchallenged as the dominant currency. With the growing importance of China and other emerging economies, the relative economic importance of the United States has diminished. After peaking in February 2002, the trade-weighted value

Figure 4.4 US trade-weighted dollar index, 1975–2014
Source: Bank of England.

of the US dollar slid lower through much of the decade, in 2007 breaking out below the channel it had occupied for the previous 30 years (Figure 4.4).

In sum, while it does appear to be the case that mineral prices have broken the trend of 30 years, the causes may not be quite those envisaged by the adherents of the super cycle. It is also questionable whether what is going on here is actually a cycle at all or a step change in the objective conditions facing the industry and prices within it. The new conditions might perhaps more usefully be represented by a horizontal line running through a new, higher, price plane (a new mean to revert to) rather than an upwards sloping trend. The factors shaping this new world may include a shift in the key driver of demand towards the emerging market economies, but they operate primarily through the supply side of the industry, through the effects on the long-run marginal costs of production, through the shift in relative prices brought about by changes in other economic sectors and through currency factors.[14]

Commodities as an investment

'It is our belief that investment funds have been the major driver behind the record high prices we have seen in many commodities in recent years, including copper, the primary raw material in our business. These high prices, which we do not believe are warranted based on fundamentals, have caused major challenges for our industry, including increases in working capital and, even more damaging, substitution. One only has to look at the copper industry to see the disconnect between fundamentals and price. In the last year, the price of copper has increased over 100 per cent, while at the same time the quantity of materials stored in warehouses has also increased substantially. Thus, while the copper market was in surplus, the price more than doubled. We believe that if investment funds were to have smaller positions in the market it would greatly reduce the volatility and bring prices more in line with fundamentals and costs.'

Such was the view expressed by Jeffrey Burghardt, spokesman for the Copper and Brass Fabricators Council, before a hearing of the Commodities Futures Trading Commission (CFTC) on 25 March 2010.[15]

At the same time that consumers and producers of mineral commodities were waking up to the transformation in their markets being wrought by China's rapid industrialisation, interest was growing in the idea of investment in commodities as a 'asset class'. Investment guru, Jim Rogers, in his book *Hot Commodities* published in 2004, argued that the history of the past 140 years showed a regular alternation between long bull markets for equities and for commodities. He believed that a combination of China's growth and the neglect of investment in new production capacity foreshadowed the beginnings of an extended bull market for commodities.[16] He had set up a commodities fund in 1998 (in his view, the bottom of the market) in order to capitalise on this. Another investment luminary, Jeremy Gratham of GMO, proclaimed in more apocalyptic terms that the 'days of abundant resources...are over forever' and that 'price pressure and shortages of resources will be a permanent feature of our lives'.[17]

The idea of commodities as an asset class was not new. The original commodity index fund, the Goldman Sachs Commodity Index (GSCI), was launched in 1991 as a means by which investors could gain exposure to a basket of commodities. The index was acquired by S&P in 2007 and became the S&P GSCI. The other industry leader, the Bloomberg Commodity Index (previously the DJ-UBS index and before that the Dow Jones AIG index), was launched in 1998, the same year that Rogers launched the Rogers International Commodity Index (RICI). These funds are essentially passive investments, which is to say, they are intended to be bought and held rather than actively traded and designed to appeal to big long-only holders of investments such as pension funds and insurance companies. The composition of these funds varies. The S&P GSCI is heavily weighted (around 70 per cent) towards energy, with metals accounting for only around 11 per cent of the fund by value. The Bloomberg Commodity Index has around 39 per cent in energy and 28 per cent in metals.[18]

Investment in these funds is effected through the purchase of commodity futures. These are rolled over by being sold at, or prior to, maturity and immediately replaced with new purchases of futures with a more distant maturity date. Contrary to what might be supposed, returns from commodity funds do not necessarily always follow commodity prices. This is because returns arise from three distinct sources: rising spot prices, the 'roll yield' that occurs in markets where futures prices are below spot prices (so-called backwardated markets) when positions are rolled over and the collateral yield, which is the interest earned on the money that has been put into a fund but which is not needed to hold a futures position (since such positions are bought on margin). Given these varied sources of returns, the performance of commodity investments can be complex. It is entirely possible for rising spot prices to be associated with diminishing fund returns and for falling spot prices to be associated with increasing returns.

Bolstering the case for commodity investments was a highly influential paper by Gary Gorton and Geert Rouwenhorst, published in 2004.[19] In this paper, Gorton and Rouwenhorst created an index of commodity futures monthly returns for the period 1959–2004. Their analysis of this index led them to conclude that the returns and risk profile of investments in commodity futures were similar to those of equities, as represented by the S&P 500, and the returns better than those for bonds. It further suggested that commodity returns were negatively correlated with equity and bond returns, a tendency 'due, in significant part, to different behavior over the business cycle'.

This last point was of considerable interest to the investment industry. For large investors having to make regular payments such as pensions, the predictability of returns is extremely important. Combined with the prospect of a prolonged bull market in commodities, the temptation was too much to resist. Money poured into commodity index funds. Standing at an estimated $9 billion at the end of 2002, by the second quarter of 2008, such commodity funds had $200 billion under management.[20]

There were other investment vehicles available for those interested in benefiting from exposure to commodities. In contrast to the long-only (buy-and-hold) index funds, many of these were long–short investments, meaning that investment managers could sell commodities short in order to benefit from falling prices as well as going long to benefit from a rise in prices. Most of these, in common with the index funds, operated though futures markets on the world major exchanges such as the LME, Comex and Nymex. The first group of these investors were commodity trading advisors (CTAs). These 'technical' investors used computer models based on past patterns of price behaviour to anticipate future price movements and to profit from them. The nature of their investment strategies meant that they were sometimes referred to as 'trend-following' or 'momentum' investors. A second important group was the hedge funds. These included both generalist hedge funds that periodically dipped into commodities and specialist commodity hedge funds. Like the CTAs, they could go long and short in the market but their investment decisions tended to be more discretionary – more judgment based. Thus, while they could, and did, use technical tools, more often the big decisions flowed from their perspective on developments in the broader economy and from their research into the fundamentals of individual commodity markets. A third group of long–short investors included the proprietary trading desks of the major investment banks that traded commodities on their own account.

A final vehicle for commodity investments was the exchange traded fund (ETF). Historically, it was difficult for the small investor to invest in precious metals such as gold. Most private investors did not have access to futures markets and did not want have to buy and store a physical commodity. The best way generally to get exposure to a commodity was thus to buy shares in a company producing the relevant commodity. An ETF offered the opportunity for an investor to buy and sell a paper investment backed by a physical commodity. The company marketing the ETF took on responsibility for acquiring and storing the metal. This had the advantage of giving the investor exposure to a metal's price without having to worry about the management risk which inevitably came with holding equity in a mining company.

The first gold-backed ETF was launched on the Australian Securities Exchange in 2003, following which numerous similar funds were launched in the major financial centres of the world, Zurich, London, Paris, New York and Mumbai. ETFs were also launched for the other precious metals, silver, platinum and palladium and for some major non-ferrous metals (although the take-up for these latter ETFs has been modest). At their peak, in December 2012, there were 2691 tonnes of gold held in ETFs, equating to over 90 per cent of gold mine production in that year.[21]

Impacts of 'financialisation'

These commodity investments were not confined to mined products but were spread across the whole spectrum of commodities, energy, metals, food products and non-food agricultural products (such as rubber and jute). The large scale of the investments gave rise to concern in the metals and mining industry, and even more so among their customers, that they distorted prices, pushing them in boom periods higher than they might otherwise have gone. Certainly, the step up in metal prices in late 2005 and early 2006 coincided with a surge of interest in commodity fund investment. There was growing talk of the 'financialisation' of commodity markets. What had previously been regarded as dull functional raw materials were becoming a plaything for portfolio managers having little regard to the uses of these raw materials or the underlying conditions of their physical markets.

When the CFTC decided to hold hearings on the matter, there were some outspoken comments from the industry, as the quotation at the start of the previous section illustrates. Those deemed by some to be investors helpfully adding liquidity to markets were deemed by others to be speculators distorting these same markets. Another straight-speaking exponent of the view that speculators were having a malign influence on the market was wealth manager Michael Masters, who told the CFTC that 'speculators far outnumber bona fide physical hedgers and therefore play the dominant role in price discovery'. Moreover, that 'when speculative euphoria takes over, speculators can and will drive prices to levels that do not reflect supply and demand conditions'.[22] The chairman of the CFTC, Gary Gensler, seemed generally sympathetic to these views, in his pre-confirmation letter, stating, 'I believe that rapid growth in commodity index funds was a contributing factor to a bubble in commodities prices that peaked in mid-2008'.[23] During Gensler's tenure as chairman, the CFTC attempted to impose limits on the number of commodity futures and swaps contracts which any company could hold in a CFTC-regulated market, although this was struck down by the courts in September 2012. A further attempt to introduce such limits was launched in November 2013.

The view that speculators were driving markets was partly circumstantial. As already indicated, the big increase in prices had coincided with the increase in investment in commodities, and all of this money had to be going somewhere, an argument sometimes referred to as the 'weight-of-money' argument. Unfortunately, these arguments were rarely, if ever, accompanied by an explanation of the mechanisms by which investment in futures flowed back into the spot and short-dated contracts which were the basis of most physical business. In times of contango markets – those in which forward prices exceed the spot price – the link is evident enough. When an investor buys a forward contract, the person taking the other side of the trade will buy spot metal and put it on deposit so as to be able to deliver the metal when the forward contract becomes due and finance this stock using the contango (the differential between the sale price of the forward and the cost of the spot purchase). The purchase will cause the spot price to rise. Often – although not necessarily always – this activity will show up in an increase in metal stocks. However, through much of the boom period, markets were in backwardation, which is to say forward prices were below spot prices. Under such circumstances, it was not at all clear how activity in the forward market would feed back into spot prices.[24]

Numerous studies have been conducted into the effects of financialisation. While a few of these – most notably a report from UNCTAD[25] – found evidence that financialisation had a material influence on the prices of traded commodities, most concluded that the use of futures market to make commodity investments has relatively little impact on commodity prices. Indeed, some provided evidence suggesting that speculative positions follow price movements rather than lead them.

In 2006, the IMF investigated the commodities boom and its causes, concluding that there 'is little support for the hypothesis that speculative activity... affects either price levels over the long run or price swings in the short run.'[26] Studies by the UK Treasury and the US Interagency Task Force on Commodity Markets in 2008 endorsed these findings as did a study commissioned by the Organisation for Economic Co-operation and Development (OECD) in 2010.[27] The IMF took a further look at the subject in 2010, once again concluding that '...there remains little evidence that financial investment has a significant sustained impact on commodity prices above and beyond current and expected supply-demand fundamentals.'[28] In short, the evidence strongly suggested that the high prices experienced during the boom years were simply a case of a rightward shifting demand curve interacting with a vertical supply curve. In these circumstances, prices will rise to whatever level is necessary to ration the available supply. The fact that many commodities, such as iron ore and the minor metals, which were not traded on futures exchanges and could not therefore be subject to the effects of financialisation, experienced extreme prices during this period would also seem to bear out the belief that the effects of speculation were marginal.

As commodity markets moved into better balance, the enthusiasm for commodity investments began to wane. Commodity funds had a difficult time in the wake of the GFC when prices fell and many markets moved into contango, depriving investors of the benefit of the roll yield. In addition, very low interest rates suppressed the collateral yield available from commodity index investments.[29] Certainly there was precious little evidence of any negative correlation between equity markets and commodity markets through this period; they went down together. The resurgence of commodity prices in 2011 temporarily revived interest in commodity investments, but thereafter commodity funds continued to struggle. Over 2011–2013, a period during which the S&P500 rose 44 per cent, the S&P GSCI Total Returns index declined at an annualised 1.4 per cent. It declined further in 2014.[30] ETFs did rather better than commodity index funds since precious metals held up better than industrial commodities over this period, but they too suffered a major sell off in 2013 as gold prices plunged. By 2012, some major institutional investors, such as the Illinois Teachers' Retirement System and the California State Teachers' Retirement Systems (CalSTRS), were beginning to go cool on commodity-based investment and reduce their holdings.[31] In November 2014, the value of commodity-based funds was 40 per cent below the April 2011 peak of $256 billion.[32] The amount of gold in ETFs had fallen by a similar proportion.

As the allure of commodity investments faded, the interest of the financial sector in commodities faded with it. Many companies offering investments in commodities were forced into closure.[33] The banks too were retreating from commodities. The US Federal Reserve, which had expressed concern about the implications for the stability of the financial system arising from the bank's exposure to commodities, moved to restrict the banks' involvement in physical trading.[34] JPMorgan, Barclays, RBS, UBS, Standard Bank and Deutsche Bank announced that they were either withdrawing or significantly reducing their direct involvement in commodities and commodity investments.[35] Opposition had also emerged among metal consumers to the banks' ownership of LME warehouses, this on the grounds that the ability of those engaged in metal trading to influence the rate of delivery from warehouses constituted a conflict of interest.[36] Under pressure from the courts and the CFTC, JPMorgan and Goldman Sachs sold off their metal warehousing businesses in 2014. Although ETFs were moving into the mainstream and becoming an established part of the fundamentals of the precious metals markets, elsewhere, a process of 'definancialisation' was setting in. Mineral commodities were less and less moving collectively in response to big investment themes. Slowly, they were being returned to the grittier realities of the physical market place and their own specific fundamentals.

Changing price dynamics – the case of iron ore

On 5 July 2009, four executives of the Anglo-Australian miner, Rio Tinto, were detained by the Shanghai state security bureau. The detainees included Stern Hu, an Australian citizen of Chinese descent, who was at the time Rio Tinto's head of iron ore sales in China. No formal announcement was made as to the reasons for the arrests of the 'Rio 4', but information leaking out from official circles suggested that the charges related to their having been in possession of state secrets. China's large steel mills had recently been subject to a settlement on annual iron ore prices which the state authorities deemed unreasonably disadvantageous. It was suggested that the iron ore companies had had the benefit in the negotiations of confidential information on the Chinese negotiating stance and that the settlement had been damaging to China's economic interests. In point of fact, the charges that were eventually brought related to the taking of bribes and stealing commercial secrets rather than to espionage. The case was heard in March 2010. It was conducted behind closed doors, and the full details of the case were not made public. All four men were convicted and sentenced to prison. The court did not reveal whether those who paid the alleged bribes had been tried or convicted.[37]

China's impact on the level of mineral prices has already been discussed at length. No less important for some commodities was its impact on the processes of mineral pricing.

The structure of Chinese industry and the manner in which it conducts business are substantially different from the structure and practices of other major mineral-importing nations in Asia. The backbone of the industrial economies of Japan and the Republic of Korea are the vast conglomerates, known respectively as *keiretsu* and *chaebol*. These are sprawling business empires, sometimes centred on a cluster of financial institutions, comprising networks of interlocking businesses and shareholdings. Some of the better-known conglomerates in Japan are Mitsui, Mitsubishi and Sumitomo and in Korea, Hyundai, Samsung and LG. These companies are highly structured and centralised organisations, which do their businesses in rigid and formal ways. This provides for stability and continuity in their business relations with customers and suppliers alike.

China's business culture is very different. The largest corporations are state owned and, like state-owned corporations everywhere, have somewhat ambiguous characters, serving as they do a mixture of commercial and political objectives. However, the most dynamic parts of the economy are the smaller private companies. These are much more obviously commercial and often more local in their scope. In contrast to the contract-based mindset of the large Japanese and Korea giants, these companies have a trading mindset. Businesses are set up quickly to capture profitable opportunities and sometimes just as quickly shut down when the opportunity passes. The nickel pig iron business discussed in Chapter 3 is a good example of this. Notwithstanding China's formally communist status, these companies are quintessentially capitalist. Business is fiercely competitive and short term in perspective. The objective is to enrich the owners quickly;it is not to build lasting industrial empires.

The trading mindset naturally shows through in China's processes of mineral procurement. Where Japanese and Korean companies sought to tie in supplies and suppliers over multi-year periods, Chinese buyers were focused on near-term price and availability. Suppliers, who had got used to working with a small number of large, technologically sophisticated customers on the basis of long-term contracts, had to adapt to a situation where they faced numerous dispersed small-scale and sometimes low-tech buyers of their products with an emphasis on the short term. The new situation called for a wholly different approach. It involved, for example, the use of smaller vessels for spot cargoes going to second- or third-tier ports; the maintenance of stockpiles at Chinese ports so as to be able to supply small, one-off, last-minute orders; the use of agents to sell products in China's labyrinthine interior; the bagging up of bulk cargoes for transport and delivery in small batches inland; and a preparedness to sell heavily discounted sub-specification materials to low-tech producers who cared more about low-cost raw materials than they did about waste levels or energy efficiency.

In addition to having to change their sales processes, mineral producers also had to adapt their sales strategies. They needed, for example, to hold more of their product back from annual contracts so as to have material available to deliver on a spot basis. And they needed to encourage the development of more sensitive indicators of spot market conditions to help guide the pricing of spot deals. To meet this need, globalCOAL was established in 2001 to provide an online trading platform for spot cargoes of thermal coal and to publish marker prices. Up to this point, effectively all seaborne thermal coal had been traded on annual contracts. In 2013, globalCOAL added a contact for metallurgical coal. Indexes of spot and forward prices were also issued on a daily basis by commercial publishers like Platts.

Nowhere were the effects of this change more evident than with seaborne iron ore, an industry dominated by the so-called Big Three producers: Rio Tinto and BHP Billiton in Australia and Vale in Brazil.

Trade in seaborne iron ore was traditionally conducted on the basis of long-term contracts with prices for contract tonnages being negotiated on an annual basis between the steel mills – either acting collectively or else through an industry 'champion' – and major iron ore producers. This pricing basis grew out of the requirement of European and Japanese steel mills for predictable supplies of raw materials of known quality and that of mine producers for predictable cash flows to get large capital-intensive export-oriented mines built and funded in the 1960s and 1970s. The setting of annual prices was a slow and formal process, generally involving several months of courteous public debate followed by gloves-off negotiations behind closed doors. However, when eventually one of the producers and one of the consumers agreed a price, the convention was that all other producers and consumers in the market adopted this as the 'benchmark' for their own contract tonnages, with appropriate adjustments for differences in grade and impurities. Although often characterised as an oligopoly – suggestive of producer pricing power – because of the concentration of supply in the seaborne market, this was an oversimplification. Demand was similarly concentrated and the market highly competitive, albeit that competition took a rather stylised form. When iron ore markets were in oversupply, as they often were, steel mills were able to pick off sellers individually by offering additional sales tonnages in return for price discounts, thereby ensuring a race to the bottom for prices and a sustained period of real-terms price decline, as described in Chapter 1.

In the mid-2000s, these arrangements began to break down. With the major producers unable to respond quickly enough to the rapid growth in demand for iron ore, a large number of small high-cost producers in China and India began to enter the market. Not being part of the annual round of benchmark price negotiations, and thus not being bound by its outcomes, many of these producers sold their iron ore on a spot basis, cargo by cargo. Given the general tightness of the market through 2005–2008 and the natural lags in the contract pricing system, these spot

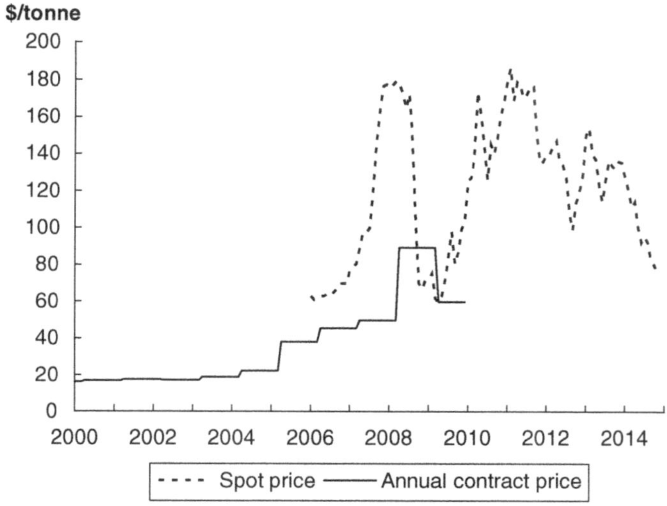

Figure 4.5 Price of seaborne iron ore, 2000–2014

Note: Contract prices are for Australian iron ore fines sold to Japan. They have been converted from basis FOB dry metric tonne unit to CFR China, $/tonne 62 per cent Fe, to make them comparable with quoted spot prices.

Source: Industry sources, Index Mundi, and author's calculations.

prices were very much higher than contract prices, as can be seen in Figure 4.5.

For a while, the large producers clung on to the contract pricing system on the grounds that it had served them well over a long period and that, with a softening of the market, spot prices would drop back below contract prices. However, with each passing year, the spot market was taking a larger and larger share of the global seaborne trade and diminishing the coverage of the benchmark system. By 2008, spot cargoes accounted for almost 40 per cent of total seaborne trade and the majority of iron ore sales in China. In 2001, the spot market was only 15 per cent of seaborne trade. Moreover, with no sign of a significant softening of the market, the opportunity cost of adhering to the contract system was becoming enormous for the Big Three.

Several factors combined to bring the old benchmark pricing system to an end. First, the GFC of 2008, and the dramatic slowing of the global economy which followed it, resulted in spot iron ore prices falling sharply, to below contract prices (see Figure 4.5). Rather than continue to buy iron ore at contract prices, some consumers simply reneged on their contractual obligations and bought spot material at the lower spot price. Producers realised that, in clinging to the benchmark system, they had in effect simply been according the steel mills a free option. It had given them no price protection whatsoever. Worse, some of their contract customers had been undermining the system (and making good profits) by selling on part of their benchmark tonnages in the spot market to smaller mills.

The second factor was the politicisation of the price negotiation process. In the wake of the GFC, the Chinese State Council had authorised the China Iron and Steel Association (CISA) to negotiate iron ore prices on behalf of the steel mills in the national interest. CISA not unnaturally adopted an extremely tough stance in the 2009 price round demanding a 40 per cent price reduction. The market, however, recovered far more quickly than anyone had thought likely and, in May 2009, with spot prices once more climbing steeply, the big Chinese steel mills followed the Japanese and Korean mills and settled a benchmark price reduction of 33 per cent independently of CISA. Inevitably, there were recriminations in China between the steel producers and CISA, but CISA was eventually forced to back down in the face of market realities. However, the damage had been done and the intervention of a state-authorised agency in what had hitherto been a commercial negotiation had generated significant distrust among the Big Three.

A third factor marking the end of the benchmark system was the case of the Rio 4 described above. On one level, this could be seen as illustrating the politicisation of the process of iron ore price negotiation. However, it had a more direct relevance to the iron ore pricing process. Some reports suggested that the Rio 4 had been taking bribes from steel mills in return for allocations of iron ore at benchmark prices, to avoid them having to pay higher spot prices. Rio Tinto said it could find no evidence in its records that this was so and claimed that those charged did not have the authority to make these decisions on their own. The Rio 4 did admit though to taking bribes from small and medium private mills in return for assisting them to obtain supplies of iron ore at better prices than they were being offered by the state-run mills, who were on-selling their contract tonnages. This might have been accomplished by tipping off the favoured mills about the likely level of winning bids in auctions for spot tonnages.[38] Either way, the case exposed the potential for corruption inherent in trying to operate simultaneously two different price systems.

By 2009, the Big Three had become less and less convinced that the benchmark pricing system was appropriate to the new market conditions. BHP Billiton in particular had come to the view that the future lay in shorter-dated pricing and had been talking about the need for changes in the pricing system since 2006. Rio Tinto had been more resistant to change but was rapidly converted by the events surrounding the Rio 4. Vale was the most reluctant to abandon the benchmark system. This was because Brazil had benefited under the benchmark system from price concessions from Asian steel producers designed to equalise the freight costs from Brazil and Australia into the Asian market and thus keep competition among producers keen. Clearly, such arrangements would be far harder, if not impossible, to sustain under a more open, market-based pricing system.

However, the writing was on the wall, and by the time negotiations for 2010 contract prices arrived, the benchmark pricing system was effectively dead. In its place, there emerged a quarterly pricing system based on published spot market price quotations. These spot market quotations are produced daily by trade publications – Platts, Metal Bulletin, The Steel Index – and are typically based on the delivery of iron ore, inclusive of costs and freight, to Chinese ports (62 per cent Fe CFR China). Since 2009, there has also been an actively traded swaps market for iron ore on the Singapore Exchange (SGX). While Indian iron ore was the basis of many of the original recorded spot deals when the market moved to short-dated pricing in 2010, Australian ore has since come to dominate the spot trade.

China's emergence as the dominant buyer in world commodity markets has thus had important implications for the manner in which mineral commodities are traded as well as the prices at which they trade. The general effect has been to extend the range of mining company customers and to reduce the time frame of contracts. There are indications that the market in copper concentrates and bauxite are following coal and iron ore towards more spot-based deals.[39] While the reasons for this trend are evident and while the new arrangements bring some benefits in terms of transparency, they also pose some challenges. Given the speed at which circumstances in China can change and the lack of transparency about what is going on in the country, the new situation adds to the difficulties that mining companies face in assessing prices for the purpose of planning their production and investments. It also has implications for their cost of capital in so far as increased price risk – or the perception of price risk – will tend to lead investors to accord mining shares a lower stock market rating.

5
An Industry Transformed

Blinking into the light

When BHP and Billiton merged in 2001, Billiton's chief financial officer, Mick Davis, found himself without a job, or at least one he wanted. So, in October that year, he joined Xstrata, a small Swiss-based company with a few zinc and ferroalloys assets in Australia and South Africa, as chief executive officer (CEO). The company was majority owned by Swiss trader Glencore International. In March 2002, Xstrata listed on the London Stock Exchange (LSE) and raised $1.4 billion of new equity. With these funds, it purchased Glencore's thermal coal assets, again mostly in Australia and South Africa.

These were not good times for mining, and Xstrata soon found itself under pressure. Profits declined in the company's first year of operation after listing, and the share price took a tumble in early 2003 after several brokers downgraded the stock. Undaunted, Mick Davis announced that he was in discussions to acquire MIM Holdings, a large, but relatively high cost, Australian mining company. In addition to bringing more thermal coal, MIM would also bring to Xstrata coking coal, copper, lead and zinc. The broking community was distinctly lukewarm on the deal but with little else going on in the sector it begrudgingly bought the company's claims that it could run the MIM operations better than existing management and that the acquisition would give Xstrata critical mass. Following another fundraising on the LSE, and with the support of the board of the target company, Xstrata acquired MIM Holdings for almost $3 billion in June 2003. One analyst wryly commented, 'With MIM, Xstrata kissed a frog, hoping it will become a princess'.[1]

Although it was far from obvious at the time, events taking place around the world at the start of the new millennium were about to transform the prospects for the mining industry. The industry had, as described in Chapter 1, entered the new era with modest expectations. Persistent oversupply and weak prices, allied with widespread public criticism, had left the industry with a bunker mentality. The corporate consolidations described in the earlier chapter were defensive moves undertaken by an industry seeking to rationalise its capacity and drive down its costs. Interest in new investment was low. China was interesting but viewed essentially as a side show, particularly in North America. The general sense was that so long as a mining company paid close attention to its costs and sweated its assets, it was just about possible to eke out a living.

These attitudes were to have important consequences for the industry's performance over the next few years. The focus on costs and capital efficiency had left the industry with little spare capacity. Data from the International Copper Study Group (ICSG) showed the world copper mining industry running at a record high of 93–94 per cent capacity utilisation over 1998–2001.[2] Equally importantly, the industry was to a large extent in the hands of conservative managers who had learnt their trade during the hard times. Project skills, which hadn't been much needed, were in short supply. While the industry was happy enough to round out and expand existing operations, it did not generally have the appetite to undertake major investments.

When market conditions began to pick up in 2003, the initial reaction was largely one of relief that the storm had passed. It was viewed as an opportunity to repair broken balance sheets and reward long-suffering investors rather than as a cue to go out and build new mines. Even as the trend persisted, managers remained cautious, and, as noted in Chapter 4, there were still many who considered it likely that 2004 represented the top of the cycle. At the company's half-year results presentation in February 2004, the CEO of BHP Billiton, Chip Goodyear, after noting the strong growth in demand from China warned that '...China will have its ups and downs. It will not be a straight line of growth up to the right.'[3] Around the same time, Paul Skinner, chairman of Rio Tinto, noted the improved condition of the company's balance sheet but cautioned 'we do not expect that Chinese demand for metals and minerals will continue to grow at the rate of the last year'. He went on to say that 'global markets are benefiting from generally improved economic performance but will continue to reflect political and other uncertainties....'[4] Not exactly a call to arms.

As prices continued to strengthen through the balance of 2004 and into 2005 and corporate profitability continued to improve, the industry began to wake up gradually to the possibility that something important was taking place. The notion of the 'super cycle' was getting a lot of news coverage, and for the first time in a generation the investment community was beginning to get excited about mining. However, companies faced a dilemma. Following years of investment neglect and suffering severe shortages of people and shovel-ready projects, they simply didn't have the ability or personnel to get projects up and running in time to capture the opportunity that was emerging. Supply in the mining industry is notoriously sticky; mines take years to build. However, if companies could not grow new mines in time to capture the upswing in prices, they could increase their production and sales of minerals by buying other mining companies. This might not boost the global supply of minerals, but it would allow individual companies to grow quickly and secure for their shareholders a bigger share of the benefits from the cyclical upswing.

This chapter examines the wave of mergers and acquisitions (M&A) that occurred in the mining industry as a result of rising commodity prices and the role played in this process by the global diversified miners (GDMs). It then looks at the activities of exploration companies and junior miners. The final two sections consider the response of the mining industry to the weaker price environment which followed the Global Financial Crisis and the secondary price peak of 2011.

Structural change

The global mining industry in 2000 was the product of a long and complex history. Although at various times in the past big metal-producing companies had sought to become miners in order to furnish supplies for their metallurgical operations, by 2000 the main players in the industry were companies interested in mining as a business in its own right. Typically, these companies were not concerned about where the mineral product was sold, nor were they interested in processing the mined product any more than was necessary to sell it. The core skills of the business were deemed to lie in identifying and acquiring high-quality mineral deposits, building large-scale capital-intensive projects to a tight budget, managing the political and environmental risks that accompany mining wherever it is undertaken and running mines efficiently. Because of the unequal distribution of minerals around the world, the industry naturally lends itself to a degree of concentration. Much of the wealth of the world's mining industry is concentrated in a relatively few, very large, long-life deposits, and getting hold of these deposits is a key to enduring business success.[5]

The other factor shaping the industry was its international character. Companies that wished to succeed in the business had to go where the minerals were. If they were not to be found in their home countries, then they had to go overseas. This focus on the global context of mining helps explain the otherwise anomalous importance to world mining of the United Kingdom, a small country with few domestic mineral resources. The United Kingdom's long history as a global hub of trade and finance has more than offset its geographic and geologic limitations and resulted in it becoming the base for some of the world's largest mining companies. It should nonetheless be noted that the 'global' character of the industry is somewhat partial. Most of the mining companies historically described as global have come from the English-speaking world, from the United States, Canada, the United Kingdom, Australia and South Africa, and had most of their assets in these areas.

Table 5.1 shows the largest mining companies in the world as of 2000. The assessment is based on the value of the companies' metal ore production in that year, expressed as a percentage of the world total. It excludes coal, uranium, oil, gas and industrial minerals. The advantage of this measure is that it gives a broad indication of the relative scale of the companies and permits the inclusion of state companies like Chilean copper producer, Codelco. The outcomes it produces differ from assessments based on market capitalisation, that is, companies' stock market values. The latter is the usual basis for such rankings used by the financial sector since the approach provides coverage of all a company's products and gives a better indication of the company's underlying economic strength. Regardless of the approach used, the top five or six companies tend to be the same, but rankings below this level vary quite a lot from year to year, subject to the relative commodity focus of the company. It will also tend to be the case that industry concentration will appear higher when based on market capitalisation than on value of production, since the widely differing quality of mineral assets means that value in the sector is more concentrated than output.

The strategy of M&A adopted by the larger mining companies to capture the benefits of the strengthening market was to bring about major structural change in the industry. It is, nonetheless, important to note that the drivers for this structural change were fundamentally different from those that had inspired the earlier phase of industry consolidation described in Chapter 1. This new wave of M&A was a strategy of expansion – a bull market strategy – focused on volume rather than cost reduction. With margins ballooning out and cash pouring into their

Table 5.1 World's largest mining companies by value of mine production of metals, 2000

Rank	Company	Country	Commodity	Share of world production %
1.	Rio Tinto	United Kingdom-Australia	Iron ore, copper, coal	5.1
2.	Anglo American	United Kingdom	PGMs, diamonds, coal	3.7
3.	Norilsk Nickel	Russia	Nickel, PGMs, copper	3.6
4.	CVRD (Vale)	Brazil	Iron ore	3.0
5.	BHP	Australia	Iron ore, copper, coal	3.0
6.	Codelco	Chile	Copper, molybdenum	2.7
7.	AngloGold	South Africa	Gold	2.1
8.	Phelps Dodge	United States	Copper, molybdenum	1.9
9.	Grupo Mexico	Mexico	Copper	1.7
10.	Inco	Canada	Nickel	1.7
11.	Newmont Mining	United States	Gold	1.6
12.	Freeport-McMoRan	United States	Copper, gold	1.6
13.	WMC Resources	Australia	Nickel, copper, alumina	1.6
14.	Noranda	Canada	Nickel, copper, zinc	1.4
15.	Placer Dome	Canada	Gold	1.2
16.	Gold Fields	South Africa	Gold	1.2
17.	Barrick Gold	Canada	Gold	1.2
18.	Gencor	South Africa	PGMs, gold	1.1
19.	Billiton	United Kingdom	Copper, alumina	1.0
20.	MIM Holdings	Australia	Coal, copper, zinc	0.9

Source: SNL Metals & Mining, Raw Materials Database.

coffers, companies felt they could take on a bit more cost if it gave them immediate access to significant additional tonnages of product. They could rationalise this on the grounds that long-run commodity prices were now projected to be higher than previously forecast. And banks and equity markets were only too pleased to assist them with the necessary finance.

The first shot in this battle of bull market M&A had been fired by industry newcomer Xstrata with its acquisition of MIM in 2003. But, in 2005, the year that the idea of the super cycle really caught fire, things began to hot up dramatically. Towards the end of 2004, Xstrata had approached the board of WMC Resources with a view to conducting a friendly takeover of the sort it had achieved with MIM. An Australian company, WMC Resources was a major world producer of nickel, copper and uranium. However, a lot had changed in market perceptions since 2003 and the board of WMC Resources rejected Xstrata's advances. Xstrata took its bid hostile with an offer made directly to shareholders in November 2004. Despite increasing its bid in early 2005, Xstrata was in March thwarted by a counter-offer from BHP Billiton, an offer which the board of WMC Resources recommended to the company's shareholders.

Undaunted by its failure over WMC Resources, Xstrata immediately turned its attention to another major nickel and copper producer, Falconbridge. Falconbridge's core assets were its nickel operations in Canada, a sector it dominated along with fellow Canadian miner, Inco, but it also had some high-quality copper assets in Chile and Peru. In the early 2000s, it was majority owned by Noranda. In 2005, following China Minmetals' abortive bid for Noranda, as discussed in Chapter 3, Noranda merged itself with Falconbridge, the combined entity taking on the Falconbridge name. Later that same year, Xstrata bought 20 per cent of the company from its largest single shareholder, Brascan. This was widely presumed to be the first step in a bid by Xstrata for the whole company. And so it proved, although the process of getting there was long and complicated, involving counter-bids from Inco and US copper producer, Phelps Dodge, and a referral to the anti-trust authorities in Europe. However, by August 2006, Xstrata had secured control of the company. It came at a price though. While Brascan's 20 per cent of the company had cost Xstrata C$28 a share, the cost of the balancing 80 per cent had been C$62.50 a share.[6]

By now the atmosphere of in the mining industry was febrile. The only question was, 'who's next?', a tendency encouraged by the banking community who were finding the industry's newly discovered enthusiasm for M&A highly lucrative. By missing out in the bid to acquire Falconbridge, Inco and Phelps Dodge had also missed out on an opportunity to put themselves beyond the reach of predators. They were perceived, in the language of the financial world, to be 'in play'. Canadian miner Teck Cominco, which had earlier considered – but backed off from – having a tilt at Falconbridge, decided that the time was ripe to acquire Inco instead and to this end launched a bid for the company in May 2006. This was countered by a superior bid from Phelps Dodge, which appeared to have won the support of the Inco board when the Brazilian major, CVRD (now Vale), swept in during August with an all-cash bid to secure the deal. Phelps Dodge, having thus failed to grow itself into the big league through M&A, subsequently agreed, in March 2007, to be taken over by fellow US copper producer Freeport-McMoRan.

Although gold was not strictly part of the super cycle story, gold prices were nonetheless rising alongside those of industrial metals and creating similar pressures for industry consolidation. Thus, in another all-North American tie up, Canada's Barrick Gold in January 2006 acquired Placer Dome. Somewhat earlier, in 2004, two of Africa's largest gold producers, AngloGold and Ashanti Goldfields, had come together to form AngloGold Ashanti.

Raising the stakes

By the middle of 2007, much of the middle ground of the mining industry had been absorbed into the larger companies. Of the 20 companies listed in Table 5.1, seven had disappeared (Billiton, Inco, MIM, Noranda (Falconbridge), Placer Dome, Phelps Dodge and WMC Resources) while another, Gencor, had unbundled itself into its constituent parts (Impala Platinum being the largest legacy company). However, commodity prices were still buoyant, companies were still hungry and investment banks were pushing for more deals. During the year, three of the largest mining companies appointed energetic young CEOs (all of them under 50 at the time of their appointments) with a remit to grow. Marius Kloppers came in at BHP Billiton, Tom Albanese at Rio Tinto and Cynthia Carroll – the first female boss at a major mining company – at Anglo American. Vale already had a sub-50-year-old at the helm, Roger Agnelli. These companies began to think bigger and to look with predatory interest at each other.

Rio Tinto had a reputation of being deeply conservative in its approach to investment, rigorous in its analysis and reluctant to pay too much for acquiring assets. It was a reputation that had served it well during the barren years when it had established itself as undisputed industry leader. But its standing aside from the M&A battles of the mid-2000s was leading some investors to ask if Rio Tinto was not missing out on a once-in-a-lifetime opportunity. BHP Billiton's acquisition of WMC Resources had catapulted it over the head of Rio Tinto to become the world's largest mining company (albeit the situation is complicated somewhat by the fact that BHP Billiton has significant oil and gas interests too). There was speculation in financial circles that BHP Billiton might try to acquire Rio Tinto. The companies' massive iron ore operations in the Pilbara region of Western Australia are adjacent to one another and there had long been talk about the cost-savings that might be achieved by bringing them together and sharing infrastructure.

In July 2007, Rio Tinto surprised the markets with much the largest acquisition to date, that of Canadian aluminium producer, Alcan. The US aluminium company Alcoa had in May of that year launched a bid for Alcan, a company which, ironically, had been broken out of Alcoa at the behest of the competition authorities in 1928. The offer, worth $27 billion, had been rejected by the board of Alcan as inadequate. When Rio Tinto entered the fray two months later, with a bid of $38 billion, the board of Alcan was happy to recommend acceptance. Rio Tinto rationalised the acquisition on the grounds that, in a world of rapidly growing demand for metals and minerals, the company wanted a holding of world-class assets in all of the three most important industrial commodities: iron ore, copper and aluminium. It already possessed world-class assets of the first two but not of the third. Rio Tinto was also of the view that power costs in China would in time have to rise, rendering production there less economic and creating an opportunity for low-cost producers outside China. Within the financial community there was a suspicion that Rio Tinto hoped the acquisition would make it immune from takeover by BHP Billiton. Although the deal would later be described by Dick Evans, former CEO of Alcan, and a major personal beneficiary of the acquisition, as one of the '... worst decisions ever, the largest metals and mining transaction in the history of the world at the high point in the commodity cycle',[7] at the time the financial community was generally supportive of the acquisition and it was approved by the company's shareholders in October that year.

Any hopes that the Alcan acquisition would put Rio Tinto beyond the reach of BHP Billiton were dashed when, the month following the Alcan deal was approved by Rio Tinto shareholders, BHP Billiton made an all-share bid for Rio Tinto. After rejection by the Rio Tinto board, BHP Billiton took its bid hostile in February 2008. Rio Tinto staged a robust defence. Because of the scale of the deal and, in particular its implications for the iron ore market, the competition authorities in Europe and elsewhere were bound to take a close interest in it, and their processes of assessment are lengthy. However, before these processes had been completed, the world had been plunged into the global financial crisis (GFC), and, in November, the deal already effectively dead, BHP Billiton withdrew its offer saying the risks to shareholder value had risen to 'an unacceptable level'. Although the still-hungry BHP Billiton bounced back in 2010 with a $39 billion hostile bid for the world's largest fertilizer company, the Potash Corporation of Saskatchewan (PCS), this too was thwarted, the federal government blocking the deal on the grounds that the acquisition was not in Canada's national interest.

The Brazilian mining major, Vale (it changed its name from CVRD in 2007), was keen to join the action with a big acquisition of its own. Throughout the decade of the 2000s, under Roger Agnelli's leadership, Vale had been weeding out non-core businesses such as steel and wood pulp while mopping up mining assets in Brazil. In a series of acquisitions, which included the companies Socoimex, Samitri and Ferteco, it had significantly expanded its iron ore interests in the early part of the decade. With the purchase of Rio Verde and the 40 per cent of Caemi, it did not already own in 2006, it gained control of over 85 per cent of Brazil's iron ore production. Its acquisition of Canico Resources in 2005 and Inco in 2006 had also turned it into one of the world's leading nickel producers.

Evidently this was not enough since, in January 2008, Vale launched a $90 billion bid in cash and shares for Xstrata. The hunter had become the hunted. In some ways, it would have been the perfect end to the Xstrata story. A company which had been formed in the depths of the recession and which had ridden the market up through a wave of audacious acquisitions had been given the opportunity to cash out near the top of the cycle. Xstrata's market capitalisation at the time of its first major fundraising in 2002 had been a modest $600 million so the offer from Vale represented a spectacular gain for shareholders, among whom of course numbered Xstrata's management. However, it was not to be. There were said to be disagreements with Xstrata's largest shareholder, Glencore, over marketing rights in the combined entity, but a sharp decline in Vale's share price meant that there were also going to be problems financing the deal. In any event, the project was abandoned in March 2008. Xstrata, still feeling the need to keep moving sought, unsuccessfully, to acquire platinum producer, Lonmin, later in the same year, and the following year proposed a merger of equals with Anglo American. Anglo American, which, like Rio Tinto, had been a latecomer to the M&A boom of the mid-2000s – it had acquired a large and expensive iron ore project, MMX Minas-Rio, in early 2008 – said it was not interested. This time, Xstrata was not in a position to take its interest hostile, although its approach did have the effect of pushing Anglo American into a major internal review.

This last phase of the mid-2000s M&A boom thus essentially ended in failure. Moreover, some of the M&A in the earlier phase, which had been successfully executed, had left companies heavily indebted at exactly the wrong point in the cycle. The game was up; momentum had been lost. Prices of commodities and of commodity stocks had been hammered by the GFC. Falling stock prices had reduced the potential for paper transactions while the debt market had dried up. In the weaker price environment now prevailing, there was little prospect of debt being paid down quickly by booming cash flows. Furthermore, many of the investment banks were themselves in deep trouble and were closing their chequebooks. Investors were left wondering what had happened to the super cycle and where their money had gone.

Even without the mega deals that failed, the mining industry in 2008, as represented in Table 5.2, still looked very different from how it did only eight years before, not just among the big players but also further down the industry rankings.

Comparison of Tables 5.2 and 5.1 reveals a number of important changes. First, and most obvious, is the loss of a number of middle-ranking single-commodity companies with Tier 1 (industry speak for top-quality) assets. Particularly notable is the disappearance of three of the world's four largest nickel producers: Inco, Noranda/Falconbridge and WMC Resources. Second is that the loss of middle-ranking companies contributed towards an increased degree of corporate concentration in the industry. Although this is not too evident from a comparison of the top 20 companies during the two years 2000 and 2008, where the share of global output grew from 41.3 per cent to 43.6 per cent, it is more apparent from a comparison of the top ten, where the aggregate share grew from 28.5 per cent to 35 per cent.

A third change is that by 2008 the regional composition of the largest companies had become broader. The representation of North America had been reduced from seven companies to five, while that from emerging markets such as Russia and Kazakhstan (to be discussed in the next chapter) had grown. Not unrelated to this, the role of the United Kingdom as a base for global mining companies had grown, as had the complexity of ownership structures. This was partly a product of the dual listing of companies such as Rio Tinto and BHP Billiton, which have primary listings in both the United Kingdom and Australia. It also reflected the arrival of emerging market companies with stock market listings in the United Kingdom but having their majority shareholders (along with most of their senior management) based elsewhere.

Table 5.2 World's largest mining companies by value of mine production of metals, 2008

Rank	Company	Country	Commodity	Share of world production %
1.	Vale (CVRD)	Brazil	Iron ore, nickel	8.0
2.	BHP Billiton	Australia-United Kingdom	Iron ore, copper, coal	5.9
3.	Rio Tinto	United Kingdom-Australia	Iron ore, copper, coal	4.6
4.	Anglo American	United Kingdom	PGMs, diamonds, coal	3.5
5.	Freeport-McMoRan	United States	Copper, gold	3.0
6.	Codelco	Chile	Copper, molybdenum	2.3
7.	Xstrata	Switzerland-United Kingdom	Coal, copper, ferroalloys	2.3
8.	Norilsk Nickel	Russia	Nickel, PGMs, copper	2.3
9.	Barrick Gold	Canada	Gold	1.9
10.	Newmont Mining	United States	Gold	1.2
11.	Grupo Mexico	Mexico	Copper	1.2
12.	AngloGold Ashanti	South Africa	Gold	1.1
13.	ArcelorMittal	Luxembourg	Iron ore	0.9
14.	Impala Platinum	South Africa	PGMs	0.8
15.	Metalloinvest	Russia	Iron ore	0.8
16.	Teck Cominco	Canada	Copper, coal, zinc	0.8
17.	ENRC	United Kingdom-Kazakhstan	Ferroalloys, iron ore	0.8
18.	Gold Fields	South Africa	Gold	0.8
19.	Evraz	United Kingdom-Russia	Iron ore	0.7
20.	Cliffs Natural Resources	United States	Iron ore	0.7

Source: SNL Metals & Mining, Raw Materials Database.

A final point is a reiteration of one made earlier, which is that the ranking in any year reflects the relative performance of individual commodities that year. In 2008, the price of iron ore was still well above historical levels. Since the largest mining companies were generally large producers of iron ore, this was bound to differentially benefit them. The inclusion of Cliffs Natural Resources, Arcelor, Evraz, Eurasian Natural Resources Corporation (ENRC) and Metalloinvest in Table 5.2 is largely a function of high iron ore prices in 2008.

Rise of the global diversified miner

A key feature of the consolidations of the mid-2000s was the emergence of the global diversified miner as a dominant model for the mining industry. Through much of the industry's history, mining companies had grown up focused largely on one particular region or one particular commodity. Yet, by 2008, according to accountancy firm PricewaterhouseCoopers (PwC), the GDMs – which is to say BHP Billiton, Rio Tinto, Vale, Anglo American and Xstrata (sometimes also referred to as the Big Five) – accounted for 50 per cent of the market capitalisation of the industry's largest 40 companies. The next largest grouping, the 14 largest gold companies, accounted for a further 26 per cent.[8]

Several explanations have been advanced to explain the success of the GDMs during these years.[9] The first and most obvious explanation was that the world economy itself was globalising. Through the 1990s and 2000s, investment had been flowing increasingly freely across national borders and supply chains were being reshaped to capture differences in regional competitive advantage. Many considered globalisation an unstoppable force, the pattern of the future.[10] This played to the strengths of companies with a global outlook and global reach. While in a sense this should have always been a natural perspective for mining companies given the broad geographic distribution of mineral resources, there had always been significant barriers to the free flow of capital in the sector, these barriers being both political and practical in nature.

The boom in commodity prices initially led to the lowering of some of the political barriers to cross-border investment as countries with mineral resources looked to overseas capital to assist with their development, a subject discussed in Chapter 7.

With regard to the practical challenges of operating globally, advances in technology were helping to dismantle some of these. Mines are highly specific with respect to their geological conditions and mining requirements. Mining companies, by comparison with manufacturing companies, generally have less scope for using standardised production processes across different geographies or for being able to gain commercial advantage by building on existing brand awareness. Accordingly, one of the biggest practical challenges faced by mining companies with global aspirations historically was to demonstrate how a large mining company could operationally be more than just a collection of separate mines.

Economies of scale at individual mine sites had been long understood. But the attainment of corporate-level 'economies of scope' – making global mining companies something more than the sum of their parts – had proved more elusive. However, things were changing. Improvements in communications, computerisation and web-based technologies not only made it feasible to manage companies globally, but large networked companies were able to achieve economies in their financial management, marketing and procurement activities and in their use of common services such as law, taxation and environmental support. It also enabled them to disseminate best practice, for example, on health and safety and community affairs, across their businesses and project a common identity globally. Most of the GDMs sought to give expression to this process by promulgating a so-called 'one company' policy, the hope being that this would give their internal organisation greater coherence but also that it would enable them to use their success in one geographical sphere to position themselves as 'developer of choice' in another.

The other thing that had changed was how mining companies were viewed in financial circles. Reference has already been made to the role played by the financial community in driving the commodity boom and in encouraging structural change in the industry. The major investment banks such as Goldman Sachs, JPMorgan, Citi and UBS were enthusiastic cheerleaders for the commodity super cycle and for commodity investments. Many of them engaged in proprietary trading in commodities in addition to trading on behalf of clients. Their promotion of mining stocks and the consequent re-rating of these stocks which took place in 2005 was a logical extension of their positive outlook on commodity markets.

However, it was a bit more than this. In addition to arguing that the prices of mineral commodities were destined to remain high for a considerable time to come, the banks argued that the scale and diversity of the large companies made them better adapted than their smaller brethren to thrive in a globalising economy and better able to take on the large high-risk high-return projects needed to meet future demand growth from emerging economies and from China. This was not an argument the GDMs and their ambitious CEOs were likely to contest. The banks' promotion of the shares of these companies, plus of course their preparedness to loan them substantial amounts of money, helped finance the wave of M&A that flowed through the industry in 2005–2008. To any suggestion that the prices the companies were paying for acquired assets appeared high, even assuming a buoyant price outlook, the banks and the companies had a ready answer. The pool of exceptional quality, long-life (Tier 1) mineral assets available for acquisition was limited – and diminishing. In order to secure the few that were not already tied up in one of the other GDMs, it was necessary to pay a premium.

The other factor favouring the GDMs was a change in the nature of the investor population. In the past, in stock markets dominated by private investors, investing in mining companies was commonly seen as providing an opportunity to make money playing the business cycle. Mining stocks are highly leveraged to the cycle and therefore have the potential to do very well in the upswing of the cycle for the investor who can time his or her investment correctly. These were not necessarily regarded as stocks to buy and hold. Indeed, it was often the higher-cost, lower-margin company that offered the greatest potential for cyclical gains. Within this context, single-commodity companies were simpler and easier to read. Essentially, this was the favoured model in North America during the glory days of the mining industry there in the 1950s and 1960s.

With the growth in the importance of large institutional investors such as pension funds, insurance companies, sovereign wealth funds and unit trust managers, the focus of investor interest shifted. These investors had to be fully invested and were looking for investments which could generate returns throughout the business cycle. Within this context, companies with a wide range of quality assets spread across a broad spectrum of countries and currencies were likely to provide for more stable earnings. In favouring these companies with their money, investors accorded them a premium over their single-commodity peers which in turn reinforced their competitive advantage by making it cheaper for them to raise capital.

The combination of these factors can be viewed as having differentially benefited the GDMs during these years. What, and who, the financial sector elects to believe in and to support is an important factor determining what goes and what doesn't in any industry, and mining is no exception. And what the financial sector elected to believe in and support at this time was the global diversified miner. Those within the financial community would no doubt argue that they were simply keying into a trend which was already there – that it reflected the new underlying realities of globalisation and the need for big new mineral projects – but there is a very real sense in which the finance sector played the midwife for the transformation of the mining industry in these years.

The industry's pioneers

It is estimated that, for every thousand mineral deposits explored, only one (0.1 per cent) will result in an economic discovery. The likelihood of the deposit turning out to be of world-class quality is nearer 0.03 per cent.[11] Which raises the question why anyone does it. The answer is quite simply that if one is skilful – or lucky – enough to turn up a commercial mineral deposit one can become very rich. Indeed, much of the value of the mining industry isn't actually created by mining at all but by the discovery and proving up of previously unknown mineral deposits. It is just that the odds of this activity paying off are very, very long – gambler's odds in fact.

In addition to its role in shaping the large-scale end of the mining industry, the financial sector also plays a critical role in shaping the performance and behaviour at the other, smaller-scale end of the mining industry, the end comprising exploration and small mining companies, collectively referred to here as 'juniors'.

Exploration companies are the pioneers of the mining industry, the equivalent of technology start-ups, and their prospects of success are similarly long. Nonetheless, the chance of turning up something big and making it rich – whether by developing a mine or, more likely, selling a discovery to a larger company – however remote, allied with the innate optimism of many in the mining sector, generally provides for a large army of explorers and a vibrant market in exploration assets. While some people are born to exploration, explorers are also recruited from among retirees of large mining companies or those 'let go' during cyclical dips. Their incentivisation commonly comes in the form of a large stack of share options in the exploration entity.

The high-risk, high-return nature of mineral exploration means that this is not generally of interest to the banking community. Exploration is normally funded by existing mining companies out of cash flow or by specialist exploration companies using equity finance raised through stock markets. Because of its dependence on equity financing, exploration spending tends to be highly sensitive to sentiment in the markets. Given the long-term nature of mineral exploration, it may seem strange that activity levels should correlate so closely with prevailing commodity prices, but it is when prices are high that the money becomes available. Whatever the condition of the market, it always helps when fundraising to have a clear and powerful theme to key into. And themes for miners do not come bigger or more powerful than that of a commodity super cycle.

The prospect of fast-growing Chinese markets was a catalyst for those wanting to set up exploration ventures and a good hook on which to hang a fundraising campaign. The presentations and chart books of exploration companies are often works of art, even if in practice they offer the investor little more than a promise. Conferences such as the Prospectors and Developers Association of Canada (PDAC) held in Toronto in March, the Diggers & Dealers conference held in Kalgoorlie, Western Australia, in August and the Mines & Money conferences held annually in London, Beijing and Hong Kong provide a teeming forum for those with exploration projects to promote to meet with financiers looking for likely projects to invest in. They are also the occasion for upbeat speeches on the outlook for the industry and the gifting of awards for exploration success.

Exploration finance is a specialist field, involving specialist investors. For historical reasons as well as rather obvious geographical ones, the Toronto Stock Exchange (TSX) (including the TSX Venture Exchange) in Canada and the Australian Stock Exchange (ASX) have traditionally dominated the market for exploration finance. The roles of New York and San Francisco, which were important in the past, have rather faded in recent years. Approximately 1600 juniors are listed on the TSX with another 1000 or so on the ASX.

Special reporting rules apply to companies seeking to raise money in these markets, particularly those relating to the reporting of reserves, a critical measure of any exploration company's success. Companies listed on the ASX, for example, are obliged to report their reserves according to a system known as Joint Ore Reserves Committee (JORC), a code devised and maintained by several professional mining industry bodies (the Minerals Council of Australia (MCA), the Australasian Institute of Mining and Metallurgy (the AusIMM), and the Australian Institute of Geoscientists (AIG)) in association with the exchange itself. Because of its rigour, the system is widely used outside Australia too. Companies listed on the TSX are required to comply with a similar set of standards known as National Instrument 43–101 (NI 43–101), these having been put in place after the Bre-X scandal touched on earlier to protect investors from unsubstantiated mineral project claims.

The third major market for mineral exploration finance is the Alternative Investment Market (AIM), part of the London Stock Exchange. Originally, established in 1995 to emulate NASDAQ's success in promoting technology companies, AIM became, somewhat ironically, a powerful magnet for exploration companies and small miners during the 2000s. Starting out in 2000 with only 17 mining and exploration companies, having a combined market capitalisation of £584 million ($872 million), by the end of 2006 AIM was home to 180 mining and exploration companies with a market capitalisation of £14,428 million ($28,220 million), this representing almost 16 per cent of the total market capitalisation of AIM-listed companies. In 2005, 66 new mining and exploration companies listed on AIM, with another 21 coming on the following year. Although fundraisings fell away with the GFC, AIM during these years established itself as a major international source of exploration and mining finance among smaller companies and mining established itself as one of the largest industry sectors on AIM.

The consequence of this increased availability of funding was a huge surge in exploration activity. The consultants SNL Metals & Mining and its predecessor organisations have for many years undertaken an annual survey of exploration spending. The results of this survey are shown in Figure 5.1. The spend follows the cycle of mineral commodity prices quite closely, rising strongly towards a peak in 2008, before regrouping in 2009 and then surging forward again.

There are three points to be made about this figure. First, it shows the high proportion of exploration accounted for by gold. The small scale of many gold deposits and the ready saleability of gold make this metal the favoured target for many juniors. Between 2010 and 2013, gold accounted for almost 50 per cent of the expenditure covered by the survey. Second, the increased value of exploration spend does not necessarily translate directly into an equivalent increase in exploration activity. The cost of undertaking exploration will have increased significantly over the period covered by the figure because of increases in wage costs and drill hire. Still less can it be assumed that increased exploration spend will correlate with exploration success. In fact, there is considerable evidence that the productivity of exploration activity (mineral discovered per dollar spent) has declined in recent years.[12] Third, the data are focused on private-sector exploration and accordingly do not take full account of exploration by state companies and other state

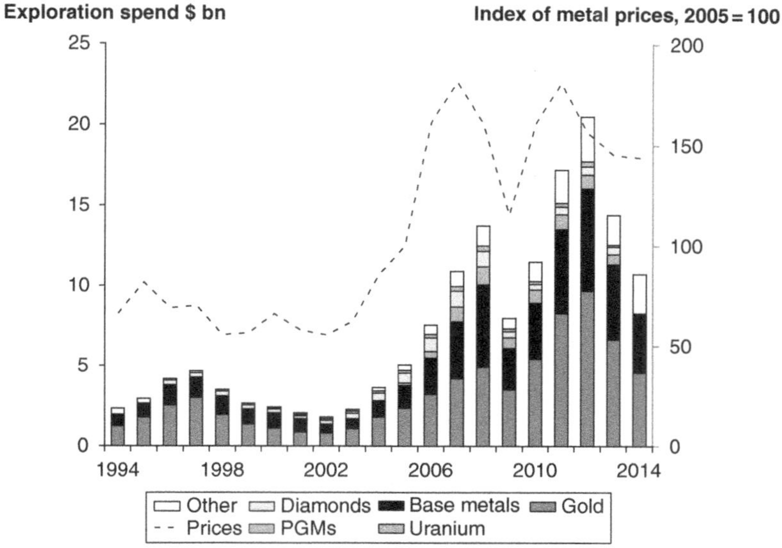

Figure 5.1 Mineral exploration expenditure by commodity, 1994–2014
Note: The survey includes uranium only from 2004. It does not include bulk commodities such as iron ore or most industrial minerals. 'Other' includes diamonds to 1995 and platinum group metals to 1999.
Source: SNL Metals & Mining and *The Economist*.

organisations. In light of the fast growth of state-funded exploration in countries such as Russia, India and above all China, in recent years, it likely that the survey data understate actual exploration spend.

The SNL Metals & Mining data also help illustrate another facet of mineral exploration, namely, how the mix of who is doing exploration changes through the cycle. Figure 5.2 shows the distribution of exploration spending in recent years split between large mining companies (majors), intermediate companies, juniors and government or other organisations. (Note that SNL's classifications do not necessarily conform to the definitions of these groupings provided in the Introduction.) It is apparent from this that spending by juniors was much more responsive to rising prices during the course of the price boom in 2004–2007 and more responsive to the falling off of prices in 2008 and 2009. Given the dependence of these companies on stock market funding, this is precisely what one would have expected. The other point illustrated by the figure is that, despite the small size of the companies in the sector, junior companies collectively account for a very high proportion of total

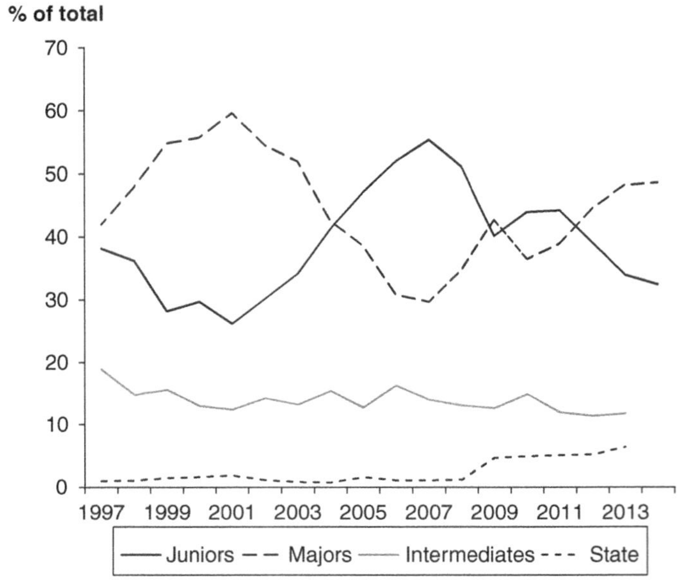

Figure 5.2 Mineral exploration expenditure by company type, 1997–2014
Source: SNL Metals & Mining.

exploration expenditures, this share rising to over 50 per cent of total spend in 2006 and 2007.

Second wind

In the title of its review of the mining industry in 2007, the accountancy firm PwC posed the question, 'As good as it gets?'[13] The question mark, it turned out, was superfluous.

The GFC, which arrived in full force following the collapse of Lehman Brothers in September 2008, brought the commodity boom to an abrupt halt. Commodity prices across the board tumbled as did the share prices of mining companies. Those worst impacted were the ones who were most indebted and whose repayments were going to be most affected by reduced cash flows. Among the GDMs, Xstrata and Rio Tinto suffered most while BHP Billiton suffered least, although even its share price halved during the second half of 2008.

However, as noted earlier, the collapse in the market and the ensuing panic turned out to be relatively short lived. Largely as a result of resurgent demand from China, commodity prices staged a remarkable recovery in the second half of 2009, and during the course of the following two years, many commodities, including some of the most important mined minerals, copper, iron ore, and gold, went on to attain new price highs.

The leaders of the mining industry, if chastened by the experience of the GFC, quickly regained their composure. Although the general euphoria which had greeted the super cycle had subsided, mining company CEOs continued to project strong long-term demand growth for the industry and talk glowingly of the transformative effects of China. Rio Tinto stuck to its claims that demand for copper, aluminium and iron ore would double over the course of the next 15 years, implying an average growth rate of nearly 5 per cent a year.[14] Indeed the evident ability of China to pull commodity markets out of the slump in 2009 in many respects served to reinforce the belief of industry leaders that China had brought about a fundamental and long-lasting change in the dynamics of their markets.

Investors, however, were not so sure. They were beginning to lose patience with expansive gestures and the alleged value-creating potential of mining acquisitions. Some of the mining assets bought with much fanfare at the top of the cycle were proving problematic. BHP Billiton was coming to the view that the Olympic Dam copper-uranium mine, the jewel in the crown of WMC Resources, was simply too big and costly to develop in the manner originally intended. Vale was experiencing large cost overruns at what had been Inco's Goro nickel project in New Caledonia and was struggling to get it into production. BHP Billiton, having the strongest balance sheet among the GDMs and still hungry for growth, significantly staged its next big acquisition outside the mining sector with the purchase of two US shale gas companies, Chesapeake Energy and Petrohawk Energy, in 2011. Other miners were also looking to the energy sector for growth. Vedanta Resources in 2011 acquired control of oil and gas company, Cairn India, while US copper producer, Freeport-McMoRan, at the end of 2012 announced that it was buying US oil company Plains Exploration & Production.

There was, however, still significant potential in the industry for organic growth – that is, growth from companies' new and existing assets – and this now became the primary focus of most companies' energies. During the boom years, while the M&A dramas were being played out, mining companies had been busy proving up projects for longer-term development, and some of these projects were now coming to fruition. Despite temporarily stalling in 2009, capital expenditure (capex) in the mining industry began once more to climb steeply and in 2011 and 2012 reached new record levels (Figure 5.3). As tends to be the case, the peak in capital expenditure lagged the peak in prices by a year,

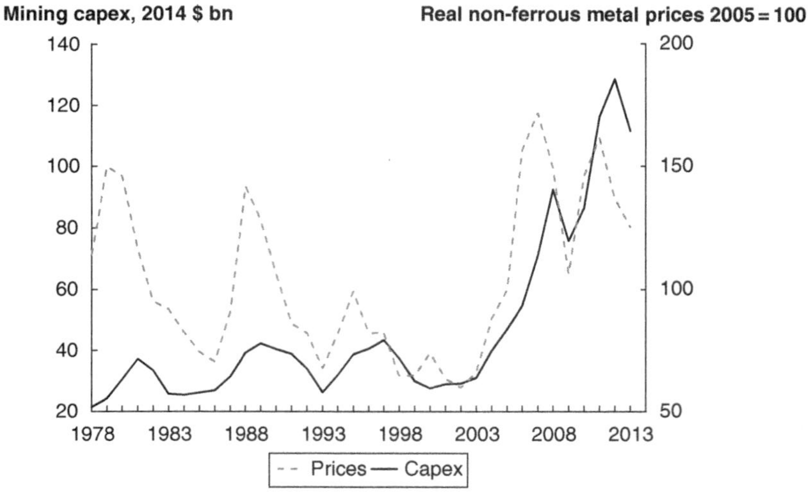

Figure 5.3 Mining investment and prices, 1978–2013
Source: Industry sources,[15] *The Economist*.

much the same pattern as that which was observable for exploration expenditure in Figure 5.1.

As a result of this investment, significant new mine capacity was coming through. The point has been made several times before that investment in mines can take many years to result in additional output. Because mineral resources are a depleting asset, some new investment is required simply to keep production where it is. Tracking back in history, it can be shown that the upturn in copper prices which took place in 1987 did not result in an acceleration in global mine production of copper until 1995, a full eight years later.[16] On this basis, the recovery of prices in 2004 might have been expected to show through in an increase in the rate of production in 2012, which is precisely what occurred. Having managed a modest growth of 1.7 per cent a year over the period 2001–2011, global copper mine production grew 4 per cent in 2012 and 8.5 per cent in 2013 according to the International Copper Study Group.[17] Chile, the world's largest copper-producing country, which had seen almost no growth in copper output for a decade, saw production grow 6.5 per cent in 2013.

In the iron ore industry, where persistent supply shortages had supported prices over a long period, something similar was taking place. As result of massive investment in Australia and Brazil, the production of seaborne iron ore surged almost 10 per cent in 2013 and was anticipated to grow a further 6 per cent a year over 2014–2016.[18]

Against the background of this high rate of investment, a softening of prices in 2012 brought about a clear shift in investor sentiment. The first of the 'end of the super cycle' research notes referred to in the previous chapter started to come out in May–June of that year. It was evident that the cost increases, which had accompanied the earlier rise in prices, were proving persistent, with the result that margins in the industry were being squeezed. PwC's annual survey of the 40 largest miners showed net profits declining 49 per cent in 2012 and a further 72 per cent in 2013. Returns on capital employed (ROCE), which had hit 22–23 per cent during the peak years of 2006, 2007 and 2011, were down to 10 per cent and 5 per cent in 2013, exactly where they were in 2002 (Figure 5.4). Worse, with global stock markets generally recovering their buoyancy, the mining sector was significantly underperforming the market as a whole, drawing investor interest and money away from the sector.

Some of the beliefs that had underpinned the notion of the super cycle were being openly challenged. Demand for mineral products may have surged as a result of China's rapid industrialisation, but it was far from obvious now that this was the start of a multi-decade shift in trend

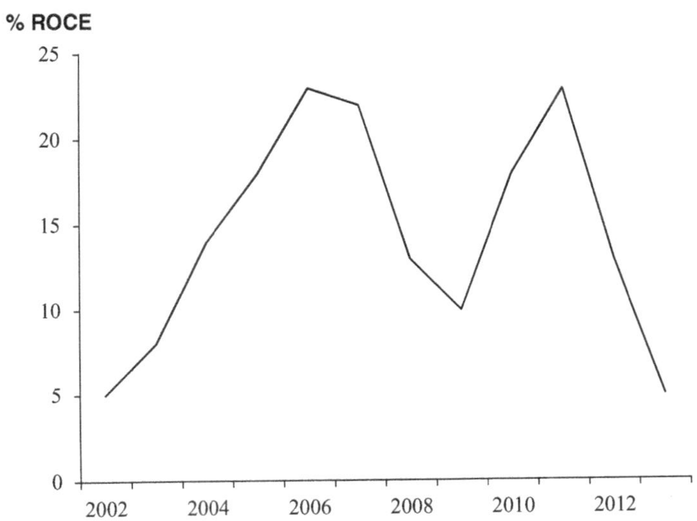

Figure 5.4 Return on capital employed: 40 largest mining companies
Source: Based on PwC (2014).

demand. Moreover, there was clear evidence that China's growth was slowing. Prices had admittedly not returned to the levels before the boom but then costs hadn't either. Without continuing evidence of strong mineral demand and strong prices, it was questionable whether it was in investors' interests for companies to continue to pour their money into creating new production capacity. It also raised questions over the logic – so seemingly plausible at the time – which had led companies to pay premium prices for Tier 1 assets during 2005–2008. Maybe this was just a normal cycle after all, albeit a big one.

Similar concerns were being voiced among investors in gold companies. After stumbling in 2009, gold prices had gone from strength to strength, bolstered by policies of monetary easing in the United States and elsewhere. But in early 2013, they took a steep dive. Unfortunately, this was not all investors in gold had to worry about. Even while gold prices had been rising, the share prices of gold companies had been lagging well behind. This turned the historical experience on its head. Half the point of investing in gold companies was that they offered a leveraged play on the gold price. When gold prices went up, gold stocks were supposed to go up more. This wasn't happening. Indeed, it appeared that the premium in the price-to-earnings ratio (the 'multiple'), which gold companies enjoyed over producers of industrial commodities, which had been a consistent feature of the mining industry over many years and a reason why gold companies generally avoided industrial commodities, was being steadily eroded. One of the factors said to be contributing to this was the growth of gold Exchange Traded Funds, which offered investors an easy and secure way of gaining an exposure to gold while not requiring them to assume the management risk inherent in investment in individual gold companies. An alternative view was that miners had simply lost control of their costs, a problem exacerbated by the fact that during the period of high prices they had invested in inherently high-cost, low-grade deposits.[19]

Back to basics

By the second half of 2012, it was apparent that the industry had moved into new territory and that another round of price increases was not on the cards. The growth model adopted by the industry during the boom years no longer applied. The game was not now about volume growth, whether through M&A or investment in new capacity, but about value. It was, as one observer put it, about optimisation and simplification, a 'retreat to the core'.[20] Adjusting to the new reality, the industry was obliged by accounting rules to write off assets which had failed to perform as expected. According to PwC, over the course of 2012 and 2013, the large mining companies wrote off $97 billion of assets.[21] While the mining industry has a long tradition of paying too much for assets and then writing them off – usually placing the blame on unlucky timing and dismissing them as non-cash items – this was value destruction on a truly epic scale.

An increasingly exercised investor community was belatedly waking up to the fact that it had been through the largest commodity boom of modern times and had remarkably little to show for it. Moreover, the investment which had been poured into new capacity in 2011 and 2012 following the secondary price peak looked as though it might be about to push mineral markets into significant surplus. The manager of one of the world's largest mining funds (conveniently overlooking the fact that such funds had played their part in stoking up expectations for the sector during the boom years) was reported as saying, '... it's incredibly frustrating for us to see companies be as reckless and profligate with shareholder capital they're meant to be custodians of.... It's happening everywhere and it's time for this to stop. We need greater capital discipline inside these companies, we need to see companies much more diligent in the analysis, we need to see boards be accountable as well as executives for these deals.'[22] Not coincidentally, late 2012 and early 2013 saw the replacement of many of the CEOs who had presided during the boom years, including the CEOs of Rio Tinto, BHP Billiton, Vale and Anglo American.

These management changes brought with them a marked shift in the language of the industry. Gone were the visionary statements and bullish claims about long-run growth in commodity demand. The new CEO of BHP Billiton, Andrew Mackenzie, sought to bring perspective to China's recent development, calling it 'a once-in-a-lifetime phenomenon'.[23] Instead, talk was of cost reductions, capital efficiency, productivity, exploitation of big data, divestment of non-core assets and, above all, shareholder value. With no immediate prospect of the market providing assistance, any growth in earnings was going to have to come from policies of self-help.[24] Many investment projects were deferred or simply cancelled as companies slashed their capex budgets. Dividends, by contrast, rose strongly – arguably too strongly, given the state of industry balance sheets[25] – as companies sought to win back the support of their disillusioned shareholders.

The one major consolidation which took place in these years involved the bringing together of Xstrata and its largest shareholder Glencore in 2013. While formally billed as a merger, the terms of the deal made it a de facto takeover of Xstrata by Glencore and resulted in Glencore personnel occupying almost all the top management posts. The merged company adopted the name Glencore in 2014. In essence, this was a bear market consolidation, although it also offered investors a new hybrid mining company model wherein physical mining assets would be used in support of the company's trading and logistics activities. Following the merger, management of the combined entity immediately set about slashing costs and cancelling projects, making clear that its focus was wholly on margins and that it was not interested in greenfield investment.

The gold sector provided rather more evidence of bear market consolidation as better-funded players used the opportunity of weak prices to rationalise industry capacity. According to SNL Metals & Mining, gold accounted for more than half the 117 M&A deals recorded in the first half of 2014, most of these occurring among the industry's smaller participants.[26]

The exploration sector was undergoing a similar reality adjustment. As 2012 progressed, funding for exploration began to dry up. By 2013, this had become a drought requiring a large number of exploration companies to suspend their activities or to close down entirely. SNL Metals & Mining survey results shown in Figure 5.1 revealed a drop in exploration expenditure of 29 per cent in 2013 and a further 25 per cent in 2014.

Despite the retrenchment in the industry which took place following the secondary price peak in 2011 and the similarities of the language of mining companies with that employed a decade earlier, this was a very different industry to the one that had gone into the boom in the early 2000s. The industry had grown massively in scale; its market capitalisation in 2013 was around four times what it had been in 2001. It had seen significant restructuring and a number of new entrants. The upsurge in exploration activity had helped to restore something of the industry's sense of adventure. London had been a particular beneficiary of these changes. From a situation where the FTSE100 contained only one mining company, it had moved to a situation where, at the peak of the boom, it contained 13 mining and metal companies and accounted for 17 per cent of the index's total value. Even following the departure of some of the smaller emerging market miners during 2013, mining was still one of the FTSE100's largest constituent sectors, with a collective market capitalisation in October 2014 of $210 billion, equivalent to 8 per cent of the market's total value.

Every bit as important as this, the industry had regained its self-respect. The industrialisation of the countries of the emerging world had put mining back on the map and given it new relevance and new purpose. Its prospects were no longer seen as constrained by the performance of a slow-growing and dematerialising West. It had become a core component of the mechanism by which millions in emerging economy countries could be lifted out of poverty and by which the demands of a rapidly expanding global middle class could be satisfied. While mining will always struggle to achieve public popularity by virtue of the physically disruptive nature of the activity, it had re-entered the consciousness of the investing public and was once again an industry in which educated young people wanted to work. However, it was also an industry that needed to adjust its strategies now that the simplistic guidance on demand and prices provided by the commodity boom had been lost, a topic to be returned to in Chapter 9.

6
Enter the Emergers

New game, new players

Norilsk Nickel is no ordinary mining company, and its incoming CEO in 2001 was no ordinary mining boss.

The mine on which Norilsk Nickel is based is arguably the richest ore body on earth. It is the source of no less than 13 economically recoverable elements ranging from nickel to tellurium. Inconveniently, it lies in the far north of Russia, at a latitude of 69.3°N, which is to say, within the Arctic Circle. Climatic conditions are extreme. There is a Polar Night of 45 days and typically over 100 days with temperatures at minus 30°C. The mining and metallurgical operations at Norilsk have their origins as a prison camp – a 'gulag' – in Stalin's Great Terror of the 1930s. Although it ceased to be a gulag following the death of Stalin in the 1950s, Norilsk nevertheless remained a closed town throughout the Soviet era, albeit a reasonably privileged one by virtue of its role as an important earner of hard currency. Its distinctive character was – and indeed still is – reinforced by the absence of any land links between the region and the rest of Russia. Everything going in and out of Norilsk has to go by sea or air. Despite these limitations, Norilsk, together with the neighbouring town of Talnakh, is home to a population of around 200,000. In1989, just ahead of the break-up of the Soviet Union, the operations at Norilsk and Talnakh were incorporated with the nickel operations of Severonickel and Pechenganickel on Russia's Kola Peninsula into the state concern, Norilsk Nickel.

Mikhail Prokhorov, the general director (CEO) of Norilsk Nickel between 2001 and 2007 was similarly distinctive. His first business venture was producing stonewashed jeans in the dying days of the Soviet

Union, but the final collapse of the Soviet Union brought with it much bigger opportunities. In the early 1990s, Prokhorov, still in his 20s, set up Onexim Bank with his business partner Vladimir Potanin. This quickly established itself as an important depository institution for the government. In 1995, with President Yeltsin's government headed for bankruptcy, Onexim Bank made a loan to the government under its 'loans-for-shares' scheme, the government's pledging its 38 per cent holding in Norilsk Nickel as collateral for the loan. (Norilsk Nickel had earlier been part-privatised with shares being distributed to its employees and sold in auction.) When in 1997 the government failed to repay the loan, control of the company passed to Prokhorov and Potanin. Given the state of the Russian economy and of the nickel market in the late 1990s, it was by no means obvious that this was such a boon. In 2001, with the company haemorrhaging money and the majority owners despairing of finding anyone they could trust with the running of the company, Mikhail Prokhorov, with no knowledge or experience of mining, decided that he would have to do the job himself. This he did until 2007 when he decided to sell his stake in the company, the stake by this time being worth several billion dollars. Not one to duck a challenge, in 2012 Prokhorov ran against Vladimir Putin in that year's presidential election.

Norilsk Nickel was just one of a number of mining companies from emerging market countries which rose to prominence during the first decade of the millennium. These companies came from all parts of the emerging world, from non-aligned countries such as Brazil, India, Chile and South Africa, from former constituent parts of the Soviet Union, such as Russia, Kazakhstan and Ukraine, and, of course, from China.

While these companies took many forms, they displayed several common features. Most of them, having their origins and assets principally in one country, had a strong national character, thus distinguishing them from the more footloose GDMs. Many of them also had relatively concentrated ownership patterns. In some cases, ownership was concentrated in the hands of a few rich individuals (oligarchs) and, in other cases, in the hands of the state. And pretty much all the companies were interested in the opportunities offered by the commodity boom and globalising capital markets to grow their businesses and extend their activities onto the world stage.

According to analysis by PricewaterhouseCoopers, the share of market capitalisation of the 40 largest mining companies accounted for by emerging market companies rose from little more than 10 per cent in 2003 to over 50 per cent in 2013 (Figure 6.1). As was revealed

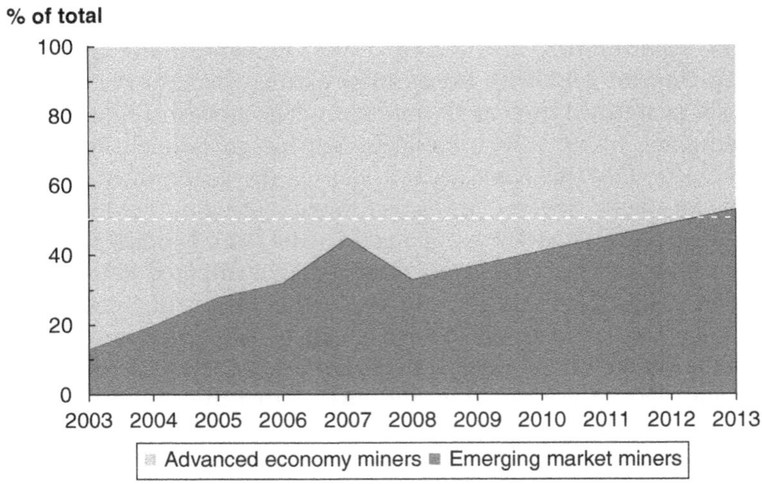

Figure 6.1 Composition of top 40 miners, 2003–2013
Source: Based on PwC (2014).

in Table 5.2, in 2008 nine of the largest 20 mining companies were from the emerging world (if one includes UK-listed Eurasian Natural Resources Corporation (ENRC) and Evraz).

This chapter looks at some of the emerging market miners and examines the reasons for their rise to prominence. It then considers the strengths and weaknesses of these companies, concluding with some observations on the issue of corporate governance.

Who are the emergers?

Companies falling under the heading of 'emerging market miner' are as varied as the countries in which they originate.[1]

The largest of them, which has already been mentioned several times, is the Brazil-based mining company, Vale. Originally founded as the state enterprise, Companhia Vale do Rio Doce (CVRD), in the 1940s, the company was controversially privatised in 1997. Its principal shareholders are large Brazilian financial institutions, but the state retains a 'golden share' permitting it to block any actions by the company deemed not to be in the national interest. In the decade that followed privatisation, the company brought most of Brazil's extensive iron ore industry under its control as well as diversifying into copper mining. It then struck out onto the world stage, acquiring the world's second largest nickel producer, Inco of Canada, and an Australian coal company, AMCI Holdings Australia, in 2007. At the same time, it changed its name to Vale to mark its new, more international, status. The company has ferroalloy operations in Europe and coal mines in Mozambique and China in addition to those in Australia.

Vedanta Resources is another emerging market miner which broke onto the scene in the first decade of the new millennium. Under the direction of its founder and executive chairman, Anil Agarwal, Vedanta consolidated into one company a number of India's major mining assets, including some which had previously been under the control of the state. The company has substantial aluminium, copper, zinc and iron ore interests in India, adding oil interests too through its acquisition of Cairn India in 2011. It also has copper operations in Zambia (acquired in 2004) and Australia. With the completion in 2011 of the acquisition of Anglo American's zinc interests, which included mines in Namibia, South Africa and Ireland, Vedanta Resources became the world's largest integrated zinc miner. The company was listed on the London Stock Exchange in 2003, although the majority of the shares in the company – nearly 70 per cent – are controlled by Mr Agarwal.

A slightly different pattern is evinced by two of Latin America's largest mining companies, Codelco of Chile and Grupo Mexico. The copper industries of both Chile and Mexico were nationalised in 1971. During the late 1980s, Grupo Mexico bought Mexico's largest copper mines out of state ownership before restructuring its assets and listing on the Mexican stock market in 1994. In 1999, it acquired a controlling interest in Southern Peru Copper Corporation from the US copper company, Asarco, and in 2009 bought the remaining assets of Asarco out of administration. The company is predominantly privately owned, with a balancing 40 per cent free float (stock market shares available to the general public). In contrast to Grupo Mexico, Codelco of Chile, the world's largest copper producer, remains firmly in state hands. Despite some tentative forays into projects overseas, it remains very much a national company in spirit and in geographic focus.

Several emerging market miners have their origins in the former Soviet bloc. When the Soviet Union collapsed in 1991, the Russian Federation declared its independence and, in so doing, set free the other constituent republics of the Union, such as Ukraine, Kazakhstan and Uzbekistan. This led to the liberalisation of many companies in the region from state control, enabling them to restructure themselves as privately owned enterprises with stock market listings. Among those companies so liberated was Russia's largest mining company, Norilsk

Nickel, described at the start of this chapter. After undergoing extensive restructuring and reform under Mikhail Prokhorov in the early 2000s, the company acquired several nickel assets in Finland, southern Africa and Australia towards the end of the decade. The company is listed on exchanges in Moscow, and its shares are traded in the form of American depositary receipts (ADRs) outside Russia. The company has two principal shareholders, Interros, the investment vehicle of Vladimir Potanin, and Rusal, the Russian aluminium company of which Oleg Deripaska is the largest shareholder, which together own almost 60 per cent of the company. The free float is approximately 30 per cent.

In Russia's iron and steel sector, most of the major players have stock market listings, although the majority of shares in these companies are, as with Norilsk Nickel, controlled by oligarchs. Several of these companies have acquired interests outside Russia. Evraz, which listed in London in 2005, acquired a range of overseas assets over the period 2005–2007 in Ukraine, Italy, the Czech Republic, the United States, Canada and South Africa. Severstal, which listed in London in 2006, acquired US steel producer Rouge Steel in 2004 and a variety of other interests in Poland, Italy, Liberia and Brazil. Russia's largest iron ore producer, Metalloinvest, has focused more on the domestic front and has not listed. The company is owned by three oligarchs, one of them, Alisher Usmanov, holding 60 per cent.

Also products of the post-Soviet privatisations are Kazakhstan's Kazakhmys and ENRC. Copper producer Kazakhmys obtained a London listing in 2005, although it remained majority-owned by its chairman, Vladimir Kim (28 per cent) and the Kazakh government (26 per cent). ENRC, which listed in London in 2007, is the world's largest producer of ferrochrome as well as a major producer of manganese, iron ore, aluminium, copper and coal. At the time of listing, 25 per cent of the company was owned by Kazakhmys and 25 per cent by the Kazakh government, in addition to having three large private shareholders. While Kazakhmys has largely confined itself to developing its business within Kazakhstan, ENRC embarked on an aggressive expansion of its activities overseas subsequent to its listing. A particular focus of its acquisitions was copper and cobalt in Zambia and DR Congo, discussed later in the chapter.

China's global miners

Chinese companies were the other large group of emerging market miners edging out onto the world stage at this time.

Chapter 3 discussed how China's need for raw materials to fuel its industrialisation was a driver for its initial wave of overseas mining investments. As noted, there was a particularly strong move from Chinese steel mills seeking to backward integrate into mining, especially in Australia, as a means to secure new supplies of iron ore.

However, China was looking for something more from its overseas mining investment than just raw materials. It wanted to develop mining companies which could compete on the world stage alongside other international miners. The Chinese authorities may have believed that, by adding to the flow of capital to the sector, its companies would boost mineral supplies generally to the benefit of all consumers (of which, of course, China was much the largest). But there was evidently also a desire to have companies which could operate profitably as miners in their own right. A presence in the global mining sector would serve to give China a window on global resource availability, on global best practice in the industry and on the global competition. In addition, it is likely that the companies headed overseas saw certain benefits for themselves, which were distinct from those of the state. As in the oil sector, where similar developments were taking place, investing overseas represented an opportunity for companies to operate more freely and expansively than was possible at home in China, as well as generating revenues and profits beyond the direct control of Beijing.[2]

Most of China's large miners are state companies, although some have set up subsidiaries with stock market listings, typically in Shanghai and Hong Kong. This not only gives the companies access to finance beyond that provided directly by the state, but it also helps educate the companies in stock market procedure, reporting and governance. Aside from the big and visible state-backed companies, there are also operating overseas, as there are in China itself, many small private Chinese mining companies. Lack of data, unfortunately, makes it difficult to assess the scale of these activities. Their production is likely to be small in volume terms but they are of significance in the mining of certain minor metals, such as cobalt in Africa.

Relationships between the state and its large mining companies are complex and the distinction between politics and commerce in this relationship blurred. State-owned enterprises (SOEs) are accountable to China's State-owned Assets Supervision and Administration Commission (SASAC), which is also responsible for appointing top executives and ensuring that major investments fall within the remit of the companies proposing them. Major overseas investment proposals have, however, also to be approved by the National Development and Reform

Commission (NDRC). The NDRC is a more overtly political body having overall responsibility for the planning of China's economic and social development. This includes ensuring that SOE overseas investments are consistent with China's national economic objectives and its foreign policy and that the company proposing the investment is the one best able to undertake it and represent China's interests abroad. Numerous other state organisations and regulatory agencies are entitled to a voice in the investment review process for foreign investments, and this before the banking institutions which will have to fund any investment and the State Administration of Foreign Exchange (SAFE) have had their say.

The complex nature of these processes was undoubtedly a factor in the poor quality of some of the investments by Chinese companies in the early stages of China's 'going out' policy and in the tendency for Chinese companies to overpay for their acquisitions. The evaluation of projects appears to have incorporated significant subjective elements and to have relied on accounting criteria rather than discounted cash flow techniques. A report on the investing activities of the sovereign wealth fund China Investment Corporation (CIC) by China's National Audit Office found evidence of 'dereliction of duty, inadequate due diligence, a lack of post-investment management and other problems.'[3] However, as noted in Chapter 4, more recently the approach has begun to change. In view of the high rate of financial losses recorded by Chinese overseas investments across all industrial sectors during the middle of the decade, the analytic rigour required in the evaluation of overseas mining project proposals and the amount of due diligence that has to be undertaken have, since 2007, been substantially increased.[4]

Given its long tradition as one of the world's largest producers of coal, it is not surprising that some of China's biggest mining companies are coal producers. In PwC's 'top 40' listing, there are several Chinese coal companies including China Coal Energy Company, China Shenhua Energy Company, Inner Mongolia Yitai Coal Company and Yanzhou Coal Mining Company.[5] While the value of these companies derives mainly from their domestic mining activities, some have ventured overseas. Yanzhou, for example, has mines in the Australian states of Queensland and New South Wales, some of these acquired as a result of the takeover of Australian company Felix Resources in 2009. China Shenhua has a project in New South Wales. China's coal companies will also be anxious to play an active part in the development of the large coal resources of neighbouring Mongolia.

Most of the overseas activity among China's mining companies, however, has been from companies with a metals focus.

The state-owned China Non-ferrous Metal Mining (Group) Co (CNMC) was an early mover in acquiring offshore assets, buying the Chambishi Copper Mine in Zambia in 1998 during that country's privatisation programme. The mine became operational in 2003. CNMC subsequently acquired additional plant and mines in Zambia and turned itself into a substantial integrated copper producer. CNMC also has the Tumurtin-Ovoo zinc mine in Mongolia and is involved in a raft of mineral development projects in other countries, including Myanmar (nickel), Laos (bauxite) and Australia (copper), as well as working as a contractor in the construction of overseas mines belonging to other companies.

The Aluminium Company of China, Chinalco, is, through its subsidiary, Chalco, China's principal producer of alumina and aluminium. Outside China it has tended to focus on other metals. As noted previously, in 2009 it became the largest shareholder in Rio Tinto and the following year became a 45 per cent partner in Rio Tinto's Simandou iron ore project in Guinea. In 2007, it acquired control of the large Toromocho copper project in Peru, which started production at the end of 2013.

China Minmetals Corporation (Minmetals) is China's largest metals trader. As noted earlier, it tried, albeit unsuccessfully, to break onto the global mining scene with a bid for Canadian miner Noranda in 2004. In 2009, it returned with a bid for OZ Minerals, a company owning several copper and zinc mines and projects in Australia. Although the bid was successful, the Australian government blocked the sale of one of the constituent assets of OZ Minerals (the Prominent Hill copper mine) on the grounds that its location was militarily sensitive. Subsequent to the purchase of OZ Minerals, Minmetals established a subsidiary, Minerals and Metals Group (MMG), to look after its international mining interests. Listed in Hong Kong (although still owned 73 per cent by Minmetals), the company is headquartered in Australia with its first CEO being an Australian national, Andrew Michelmore, formerly CEO of WMC Resources. In 2011, MMG successfully bid for Anvil Mining, which owned the Kinsevere copper-cobalt mine in DRC Congo. In 2014, it led a Chinese consortium, which acquired from Glencore the large Las Bambas copper project in Peru.

Reasons for rise to prominence

The rise to prominence of the emerging market miners during the first decade of the millennium was the result of several interlinked developments.

The first was the widespread disengagement of the state from the mining industry, which began during the liberalisations and privatisations of the 1980s but which accelerated with the collapse of the Soviet Union in the early 1990s. While under state control, mining companies in emerging economies were effectively condemned to slow, incremental growth by their lack of access to capital and skills and their lack of knowledge and experience of the industry beyond their borders. The disbanding of state controls opened the door to an inflow of entrepreneurial talent and a more commercial and expansive set of objectives, including the possibility for the first time of operating overseas.

Second is the simple fact that many of the world's undeveloped resources are found in emerging economy countries. The picture varies across commodities but, according to data published by the US Geological Survey, roughly three-quarters of known reserves lie in emerging economy countries (Figure 6.2), a finding which is not particularly surprising given that these same countries account for a similar proportion of the planet's land area.[6]

A third factor assisting the growth of these companies was financial. The commodity boom between 2003 and 2008 gave companies in the sector strong cash flows and healthy balance sheets. Coupled

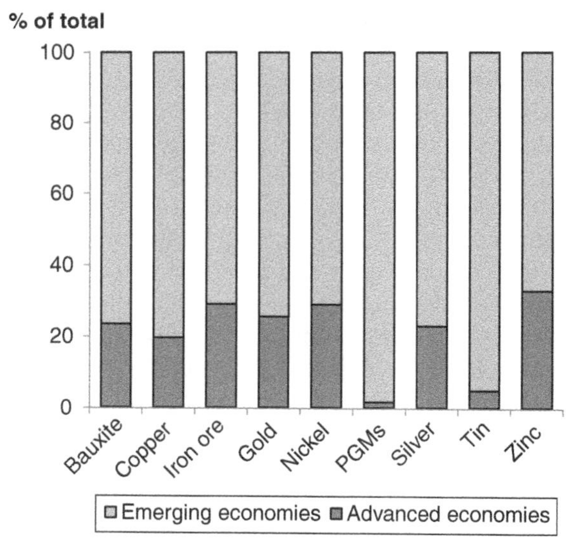

Figure 6.2 Global distribution of mineral reserves
Source: USGS (2014).

with improvements in financial reporting, this massively increased their access to capital markets. The majority of the fundraising was in the form of bank borrowing and bond issuance. Some companies, however, tapped equity markets. With the Sarbanes-Oxley Act of 2002 having imposed a discouragingly severe regulatory regime on companies listing in the United States, much of the interest of mining companies seeking a stock market listing was focused on London.

As noted above, Vedanta Resources, Kazakhmys and ENRC all did listings and fundraisings on the London Stock Exchange between 2003 and 2007, as did two Russian steel companies (Evraz and Severstal) and one Ukrainian company (Ferrexpo). Other companies listing on the LSE around this time were Peruvian silver producer Hochschild (2006), Mexican silver producer Fresnillo (2008) and Russian precious metals producer Polymetal (2011). The maturing of the banking systems and stock markets of many emerging economies, combined in some cases with high levels of domestic savings and low borrowing costs, further assisted in increasing the availability of funding for resource projects in emerging economies, most notably for smaller and mid-cap companies. Hong Kong emerged as a significant source of equity finance for mining in Asia.

A fourth factor driving growth in these companies was a quest for geographic diversification. Many of the larger mining and metal companies in emerging economies saw globalisation as a means to improve their risk profiles by spreading their assets more widely. In pursuit of this, they sometimes had the tacit support of their governments for whom the idea of hosting a global champion often appealed. The high prices which many of these companies paid for their acquisitions suggests a belief that they were justified in paying a premium to the extent that the consequent improvement in the company's risk profile would be reflected in a higher valuation of the business overall. A more diverse and global profile would also give companies a bigger and more cosmopolitan pool of management talent from which to draw, a critical factor both to furthering their global growth and making the process self-sustaining.

A fifth and final factor is that emerging market economies are the fastest-growing markets for mineral commodities. Although much of the interest in mining overseas from emerging market economies has been motivated, as discussed above, by an ambition to build mining businesses, in a few cases it has been driven by the need to secure raw materials for domestic metallurgical operations. This applies particularly to China and its iron ore investment in Australia but is visible in a few places elsewhere. Russian Aluminium (Rusal), which has a large amount of low-cost aluminium smelting capacity in Russia, but an insufficiency of local raw materials, has acquired bauxite and alumina production facilities in Guinea, Jamaica, Guyana, Ireland, Sardinia and Australia.

March of the emergers

One of the strategic ambitions of the London-listed Kazakh miner ENRC was to develop a large position in the copper and cobalt industries on the African copper belt. In August 2010, the company announced it had acquired a controlling shareholding in Camrose Resources. Camrose, a company controlled by Israeli businessman Dan Gertler, owned five copper-cobalt assets in DR Congo. What made this acquisition of interest was that these same assets had only the previous year been seized by the government of the DRC from the project developer, Canadian-listed First Quantum Minerals (FQM), on the grounds that the company was unreasonably resisting government demands to renegotiate the terms and conditions of its mining licences. The assets were then on-sold by the DRC Government to Camrose. The $175 million ENRC paid to acquire a controlling interest in Camrose appeared modest in light of the fact that FQM had spent $750 million on developing just one of the assets being acquired, the Kolwezi copper-cobalt mine. ENRC claimed that whatever the rights and wrongs of the actions of the government of the DRC, it (ENRC) had acquired title to these assets in a wholly legitimate manner.

Despite the protests of FQM – who, with their partner, the International Finance Corporation (IFC), launched a process of arbitration under the facilities of the International Chamber of Commerce International Court of Arbitration in Paris – the acquisition was allowed to stand. ENRC initially presented a united front to shareholders and the public. However, it soon became apparent that these events had sparked tensions within the boardroom. A particular source of friction was the role played in the company's affairs by its three founders who were also its largest shareholders.

When ENRC listed on the London Stock Exchange in 2007, the free float of shares (18.4 per cent) was below the level normally required for a full listing (25 per cent). However, on the understanding that future fundraisings by the company would broaden the shareholder base and that the three founders would not interfere in the running of the company, the listing was allowed to proceed. By June 2011, tensions on the board had boiled over and two of the independent directors were dismissed, the first board sacking at a FTSE100 company in more than ten years. In his valedictory letter, one of the sacked directors, Ken Olisa, wrote, 'It has become more and more apparent that the founders' influence was less transparent than is ideal; manifesting itself, for example, in the chairman frequently playing the part of the founders' messenger... rather than as leader of an independent board'. The manner of his dismissal he memorably described as 'more Soviet than City'.[7]

The increasing scale and reach of emerging market mining companies were revealing both the strengths and the weaknesses of these companies. As noted, emerging market miners are a diverse grouping, and not all strengths and weaknesses are common to all companies, but some generalisation seems warranted.

The first, and most obvious, competitive advantage enjoyed by many emerging miners is that they are commonly based in resource-rich countries. They often have privileged access – the 'inside track' – to these resources. This privilege may not be formalised but arises naturally from the practical reality that the managements of these companies are typically well connected politically and bureaucratically within the countries in which they are domiciled, understand the regulatory regime and how to operate within it, and have good knowledge of local resource development opportunities. This gives them a powerful competitive advantage relative to incomers to the country who, by definition, lack these things, as well as giving them a powerful lever in any negotiations where they need to co-opt the involvement of incoming foreign investors with technologies and managerial skills they lack. When Rio Tinto and BHP Billiton determined in 2006 that they wanted to mount exploration programmes in Russia, it was considered judicious to do so in joint ventures with local miner Norilsk Nickel.

A second advantage is that they often have a different risk appetite with respect to investing in other emerging market economies. For the diversified mining majors with Anglo-Saxon traditions, there is a strong predisposition in favour of clear-cut, formal deals, supported by legal contracts enforceable in the courts. Often the deals on offer from mineral-rich host countries fall short of these exacting requirements. They may be as much built around relationships as they are around contract law. This is something that investors based out of emerging markets may be more comfortable with and be better able to manage than are Western majors. Perceptions of risk are critical. A potential investor sitting in London and looking at South Africa may see first and foremost the challenges of black empowerment, street violence and power shortages. Looked at from Astana in Kazakhstan, the investor may see instead a country offering political stability, good banks and a sound legal system.

Closely related to this issue of risk perception is the matter of nationalism. As will be discussed at greater length in the next chapter, nationalism in economic affairs has for several years been in the ascendancy. Within this context, mining champions serving as vehicles for domestic national interests may more readily find a common, pragmatic language with resource-rich host countries which are similarly pursuing national economic and development interests. Such a perspective potentially makes the incoming investor more disposed to adopt a policy of non-interference in local affairs. Indeed, to do otherwise may be seen as an infringement of sovereignty. China sees Africa essentially as a business opportunity and as means to address its (China's) raw material requirements. It is not in Africa to reform it. ENRC's incursions into DR Congo displayed a similarly pragmatic approach.

Reinforcing the more national perspective on resources and their development is the prominent role played by the state in many emerging market miners. Effectively all the large Chinese companies investing overseas in mining are either state owned or state controlled. The state retains a 12 per cent shareholding in Kazakh miner ENRC. The state's holding in Norilsk Nickel is only small and held through a state bank, but the company publicly boasts of it close relations with the state, and its last two CEOs have been former government ministers. The state also has only a small (5 per cent) shareholding in Vale, although, critically, it holds a golden share, which it could use in the event that the company was threatened with takeover by a foreign company. The involvement of the state in these companies arguably makes them better adapted than their publicly owned competitors to engage with host nations where the state takes a similarly activist role in industrial matters. It also means that these companies can, for all practical purposes, be considered takeover proof by foreign entities. There is some evidence that companies in the resources sector with mixed state and public shareholdings and which benefit from government support and state guaranteed loans, financially outperform the wholly publicly owned competition.[8]

Close association with the state means that emerging market mining companies can often look to their parent governments, either directly or through state bodies such as banks and development agencies, to support them in their overseas activities, something on which the more stateless, transnational mining majors cannot rely. In some instances, this support may be no more than a supportive visit from a political leader or a general framework agreement affirming the intention of the two countries to co-operate in mineral development projects. However, in other cases, government-to-government agreements involving aid and infrastructure have laid the basis for a country's companies acquiring resources within a country and the licence to develop them, as, for example, with the Sino-Congolese Cooperation Agreement of 2007 described in Chapter 3. It can be, and has been, argued that in certain parts of the world this serves to tilt the playing field unfairly towards China and Chinese companies.[9] Certainly, there is a prospect that the involvement of the state in an overseas mining venture will reduce the risk of expropriation or abuse by the host nation. This was perhaps part of the thinking behind Rio Tinto bringing Chinalco into its Simandou project in Guinea.

The final point of competitive strength for the emerging market miners – albeit a slightly doubled-edged one – is their ownership structures. Most of the companies under discussion here have highly concentrated ownerships. This reduces the number of people whose views need to be taken into account when making major business decisions and accordingly enables companies to move quickly when there are deals to be done. Where the state in heavily involved, as with the Chinese companies, this may not make for fast decision-making but it does mean that effectively no consultation outside government is required. Nor, for that matter, is there likely to be a requirement to take account of domestic public or press opinion. If a project is deemed by the authorities to be in the public interest, it is set to go. The downside of this, of course, is that the ability to make quick decisions is no guarantee that the decisions will be good ones. While the submission of proposals for consultation and agreement by independently minded banks, shareholders and other stakeholders is likely to complicate matters, and mean decisions take longer to make, it may well result in better decisions and longer-lasting, more robust projects.

A raft of challenges

If changes in the global economic and political environment have accorded emerging market miners certain advantages, these companies also have their share of challenges. Some of them have already been alluded to. Many of these companies are relatively young (at least in their current forms), and their growth through the first decade of the millennium occurred in what were relatively benign conditions for the mining industry.

In common with the investments made by their advanced economy counterparts, some of the mergers and acquisitions (M&A) undertaken by emerging market companies during the boom years turned out to be costly. Norilsk Nickel's $6.3 billion acquisition of Canadian-listed LionOre Mining International in 2007, achieved after a bidding war with Xstrata (and at the time the biggest ever overseas acquisition by a Russian company), had subsequently to be largely written off. Evraz had to bear several years of heavy losses at Highveld Steel & Vanadium, which it acquired for $700 million in 2007, before selling it in 2014. Severstal has divested the US steel assets it acquired during the boom years. Vale suffered large cost overruns and technical problems at the Goro mine (now the VNC mine) in New Caledonia, acquired with Inco in 2006. However, beyond these cyclical business challenges, emerging market companies also face some more deep-rooted institutional issues.

A challenge for all companies working overseas is the need to respond constructively to the working environment and cultural norms of the host nation. Most of the GDMs have a long history of operating overseas. London-based Rio Tinto was conceived as an international investor; its first investment (from which it takes its name) was in Spain in 1873. They have also to some extent mitigated the problem of managing the difficulties of working across different business cultures by focusing their investment activities on countries sharing similar business traditions, most notably the United States, Canada, Australia and South Africa.

For companies new to overseas investment, particularly those having strong national characters and which regard themselves in some sense as national champions, the challenge is considerable. Vale struggled with the integration of Inco and its employees after acquiring the company in 2007. Having laid off almost all of the company's head office staff in Toronto and, in so doing, losing a lot of valuable experience, the company then suffered a yearlong strike of workers at its Sudbury nickel operations starting July 2009.[10] Norilsk Nickel faced similar managerial challenges at its mines in Africa and Australia, which came as part of its LionOre purchase. Both companies have also faced fierce debate within their own countries around the very principle of expansion overseas, some critics seeing this as effectively the exporting of jobs and arguing that the money spent overseas would be better used in developing domestic resources. Reportedly this was a key factor in the enforced change of chief executive at Vale in April 2011.[11] Norilsk Nickel has latterly refocused the company on its Russian assets and divested its mines in Africa and Australia.[12]

A related challenge is that of management depth and of the need when globalising for management with broad-ranging, international technical and commercial expertise. Attempts by emerging market miners to address this challenge and at the same time to bridge the business culture divide, by hiring in internationally experienced staff to help with the transition to become global players, can prove problematic. Many countries have distinctive business cultures, which are not always easy for outsiders to adjust to, and in some cases there are language issues to be surmounted. Nor can it be ruled out that highly paid outsiders coming into a company on terms substantially different from locals employees will not lead to resentment within the organisation and inhibit integration and effective working.

Chinese companies have faced significant management problems at their operations in Zambia, DR Congo, Zimbabwe and Peru, a situation undoubtedly aggravated by the practice of bringing into these countries large numbers of ethnic Chinese to build and manage the operations and the influx of small-time Chinese traders who tend to follow these migrations. In Zambia, the presence of China as an investor in the country at all is a major political issue.[13] In Brazil also, there have been expressions of concern about the scale of China's investment in the country's resource sector. In 2010, China accounted for $17 billion of Brazil's $48 billion of inward-flowing direct investment.[14] In addition to these concerns over the scale of China's inward investment in resources and their impact on community relations, there have also been concerns of a more strategic nature. These were effectively the grounds for the Australian government in 2009 blocking China Minmetals's attempt to acquire the Prominent Hill copper mine and CNMC's attempt to acquire rare earths producer, Lynas Corp, in the same year.

The backlash against China and China's business practices, and the more general tendency to nationalism in the resources sector (from which emerging market companies are potentially as exposed as the diversified majors), may be an additional factor contributing to China modifying its stance on the acquisition of overseas mining assets, as discussed in Chapter 3. To benefit from mining and secure mine offtake, it is not always necessary to own mines outright. Sometimes it is better to leave mining in the hands of companies that are more familiar with local conditions and regulations and instead develop partnerships and long-term finance and co-operation agreements with them.[15] Assistance with the creation of infrastructure also can be a powerful factor facilitating the development of mining. China is looking to such a policy in relation to the development of Indonesia's coal resources, using its sovereign wealth fund, the CIC.

Living by the rules

Finally, there is the thorny issue of governance. Concentrated ownership structures may make for fast decision-making, but there are risks involved where a company lacks appropriate checks and balances in its decision-making processes, especially when that company is listed on a stock exchange. Investors and the companies they are invested in have a two-way relationship. In return for the money they put into a company, investors have expectations about how their money will be used and how the company will behave. In companies where share ownership is divided between a handful of very large shareholders and a lot of very small ones (minorities), it cannot necessarily be assumed that their interests and perspectives will always coincide. There may be differences, for example, with respect to how much of the net revenues generated by the company should be retained within the business and how much paid out in dividends.

When times are good and there is something for everyone, these differences may be relatively easily managed. However, adverse market conditions force companies to make decisions on business priorities which can drive a wedge between the interests of majority and minority shareholders.[16] In such conflicts, there can only be one winner. The interests of the minorities will tend to be brushed aside, as they were, for example, during the boardroom battles at Norilsk Nickel in the second half of 2008 when a tussle for the control of the company between Interros and Rusal resulted in a victory for Interros and the squeezing out of the independent directors whose job it was to look after the interests of the minority shareholders. For the views of small investors to matter to a company's board, it has to need their money. In companies dominated by oligarchs and largely financed by debt, they generally don't.

As commodity markets have moved into more turbulent waters, the challenges of governance have been magnified. Investors facing poor or declining returns are more disposed to question the way in which a company is being run and way in which major decisions about its future are made.

Two cases in particular focused attention on the shortcomings of governance in the sector and led to demands among investors to tighten stock market listing requirements. The suggestion was that in its enthusiasm to establish itself as the stock market of choice for large resource companies from the emerging world, the London Stock Exchange unduly relaxed its standards.[17]

The first case was that of ENRC, already alluded to. Recognising that lack of faith in its governance was doing damage to its share price, the management of ENRC made serious efforts to improve its procedures and transparency in the wake of the sacking of the two independent directors. It also agreed in January 2012 to pay $1.25 billion to FQM to settle the dispute over the copper-cobalt assets it had acquired in DR Congo. However, it had all come a bit late. Faced with an enquiry by the UK Listing Authority into the Camrose deal and the launch of a probe by the UK Serious Fraud Office into allegations of corruption, bribery and fraud in the company's dealings, a decision was made by the three founders of ENRC in association with the Kazakh Government to buy out the company's minority shareholders and take the company private in the second half of 2013. At the same time, they bought out the shareholding of the other LSE-listed Kazakh miner, Kazakhmys. These buyouts, however, only came at the cost of adding substantially to the company's indebtedness. In a situation rich in irony, it was being reported in 2014 that ENRC was contemplating the sale of its international assets, including its coveted Africa assets to pay down its debts and refocus the company on its operations in Kazakhstan.[18] Subsequent to ENRC's delisting, the other London-listed Kazakh miner, Kazakhmys, took the decision to return some of its older, higher-cost assets in Kazakhstan to private ownership, the remaining London-listed vehicle, with a majority free-float, being rechristened KAZ Minerals.

The other highly publicised case giving rise to concerns over governance standards in the sector was that of Bumi, for a brief time one of the largest mining companies listed on the LSE. Bumi PLC was created in 2011 out of a combination of a UK investment vehicle headed by Nathaniel Rothschild and an Indonesian thermal coal company controlled by the Bakrie family. In late 2012, the company announced it was investigating possible financial irregularities at its Indonesian operations. This investigation uncovered more than $200 million of spending 'with no clear business purpose'. After falling 60 per cent below their listing price, Bumi's shares were suspended following problems in finalising its 2012 annual accounts. Attempts to recover the missing monies and to tighten corporate procedures led to a breakdown in relations between Rothschild and the Bakries, and in March 2014 the company was dissolved, with the Bakries buying back their coal assets and the rump of the company being rechristened Asia Resource Minerals.[19]

As with their counterparts from the advanced economies, emerging market miners have had mixed experiences since the passing of the commodity boom and have had to adapt to a new set of market conditions. Smaller profits have led them to focus on reducing costs and to divesting or closing some of the more indifferent assets they acquired in better times. In some cases, the adjustment has led companies to focus more on opportunities in their home countries and less on developments overseas. Those companies that have elected to maintain their stock market listings have been making adjustments to meet new, tighter governance standards (e.g., restrictions on the involvement of 'controlling shareholders' in the management of companies) while others, such as ENRC, have opted to return to private ownership. Those driven by a more strategic imperative to globalise, as with the Chinese state companies, will, it may be assumed, not be too much affected by transient market setbacks.

But if some adjustments have been, and will be, necessary, there is little likelihood that the passing of the boom which helped bring emerging market miners to prominence will see them slip back into obscurity. In fact, as demonstrated in this chapter, these companies are continuing to grow and appear generally quite well adapted to operate within the geopolitics of the new world order. They look set to be an important feature of the global mining scene in years to come even if questions remain about the exact form these companies will take and the strategies they will adopt.

7
Dividing the Spoils

Beneficiaries of the boom

'The mining boom that got underway in the middle of the last decade has already made us the richest citizens in Australian history. The extra income amounts to about 15 per cent of our economy, or about $190 billion each year. It is as though we inherited the entire economy of New Zealand in one fell swoop – and if prices remain steady, we will keep inheriting it year in, year out. In fact, with iron ore exports set to double, and LNG exports on track to quadruple, we'll probably inherit two New Zealands before long'.[1]

The commodity boom which transformed the mining industry also transformed the fortunes of mineral-producing countries. As the above quote from Paul Cleary's 2011 book *Too Much Luck* graphically illustrates, money poured into mining countries as price rose and companies scrambled to invest. The boom reshaped expectations of, and attitudes towards, the mining sector. It also created a unique opportunity for mineral-rich countries to boost public spending and promote their long-term economic development.

The mineral endowment of the world is unevenly distributed and so, therefore, is mineral production. Table 7.1 shows the largest five producers in 2013 of some of the economically most important minerals. Unsurprisingly, the world's largest countries, China, Russia, the United States, Australia, India, Brazil and Canada feature prominently. The table also illustrates the geographic concentration of production. Mineral occurrences are not spread randomly but tend to cluster. For the commodities listed, the top five producers typically account for between half and three quarters of total global production.

Because of the inertia of the sector – mines take a long time to build, and many stay operating for several decades – the picture on production

Table 7.1 World mine production of selected mineral commodities in 2013

Bauxite		
	Million t	**% share**
Australia	81.1	29
Indonesia	54.2	19
China	44.1	16
Brazil	32.5	11
India	19.2	7
Top 5	231.1	82
World	283.3	100
Coal		
	Million t	**% share**
China	3561	46
United States	904	12
India	613	8
Indonesia	489	6
Australia	459	5
Top 5	6026	77
World	7823	100
Copper		
	000 t	**% share**
Chile	5776	32
China	1707	9
Peru	1376	8
United States	1255	7
Australia	990	5
Top 5	11104	61
World	18294	100
Gold		
	Tonnes	**% share**
China	438.2	14
Australia	266.1	9
Russia	248.8	8
United States	228.9	8
South Africa	174.2	6
Top 5	1356.2	45
World	3022.1	100

Iron Ore

	Million t	% share
Australia	609	32
Brazil	367	19
China	269	14
India	125	6
Russia	106	5
Top 5	1476	76
World	1930	100

Nickel

	000 t	% share
Indonesia	811.5	32
Philippines	315.6	13
Russia	276.0	11
Australia	234.0	9
Canada	223.3	9
Top 5	1860.4	74
World	2512.9	100

Source: GFMS, RMD (SNL), UNCTAD, WBMS, WCA.

does not change radically from year to year. (The partial exception to this is China, which, as discussed in Chapter 3, has seen unusually rapid growth in the sector in recent years.) However, change does occur at the margins of the industry as mines close through depletion and as investment in the sector is drawn to particular countries because of their geological potential and, no less importantly, the attractiveness of their investment climate.

Standardised data for international flows of investment in mining do not exist, but data on exploration provide an indication of where companies are directing their interest and their dollars. The data set from SNL Metals & Mining used in Chapter 4 to show flows of exploration spending by commodity is used in Figure 7.1 to show exploration by region. Exploration activity rises and falls with commodity prices as observed previously, but over the period as a whole there was a strong increase of exploration spending across the board. The United States, Australia and Canada collectively saw growth of 450 per cent between the trough of the cycle in 2002 and 2013, and at the end of the period accounted for around a third of total expenditure. The principal gainers

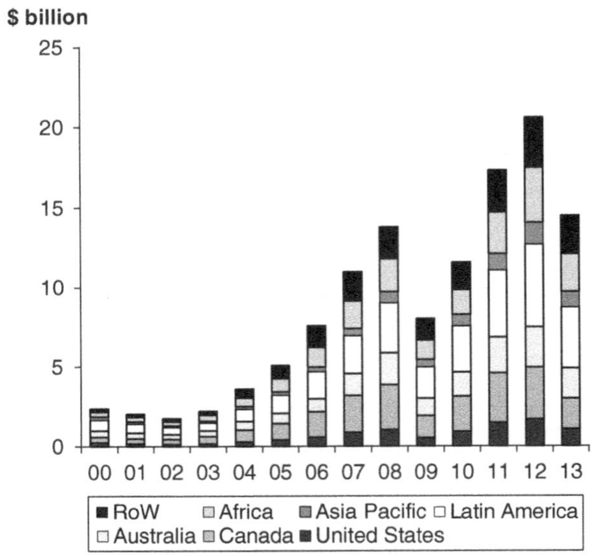

Figure 7.1 Mineral exploration expenditure by region, 2000–2013
Source: SNL Metals & Mining.

from the boom, however, lay outside these regions, among the emerging economies. Exploration expenditure in Latin America (which had been the fastest growing region in the 1990s) grew 660 per cent over 2002–2013. It was up 725 per cent in Africa, 940 per cent in Asia Pacific and over 1000 per cent in the rest of the world, a grouping dominated by China and Russia, but which also includes countries of Europe, Central and Southern Asia and the Middle East. In view of the data's partial coverage of exploration in China, it is quite possible that the indicated growth for the rest for the world is understated. What is not in question is that the growth in exploration spend was heavily focused on the emerging world.

This chapter considers the role that mining plays in the economies of mineral-producing countries and how the commodity boom sharpened interest in these countries in how to maximise their benefits from mining. It then looks at the measures adopted by governments of mineral-producing countries to increase their take from the sector and to gain greater control over the development of mineral activities within their territories, a tendency sometimes termed 'resource nationalism'. It concludes with a discussion of how global mining companies have responded to these pressures.

Mining's contribution – and mineral dependency

The size of a country's mining output does not, of course, provide an indication of the relative importance of mining to that country's economy. While the United States is a major mining country, the contribution of the mining sector is swamped by those of the much larger service and manufacturing sectors. Mining's share of gross domestic product (GDP) is 0.6 per cent.[2] For Canada and Australia, large mineral-rich countries with smaller populations and less extensive manufacturing sectors, mining is rather more important to economic life. In Canada, the contribution of mining to GDP is a little over 1 per cent.[3] In Australia, it is around 7 per cent,[4] while output from the large mining equipment, technology and services (METS) sector contributes almost as much again.[5] For some emerging market economies, such as Indonesia and Guinea, mining's contribution to GDP exceeds 10 per cent.

The importance, and potential importance, of mining to emerging economies cannot necessarily be captured wholly effectively by GDP measures. For many such countries, mining's direct contribution of GDP will only be in the range 3–10.[6] However, this understates its significance. For countries without established manufacturing or services sectors or the obvious potential to develop them, mining can be extremely important as a source of foreign exchange, a magnet for investment, a contributor to government revenues, a turnkey for infrastructure and regional development, a provider of direct and indirect employment and a catalyst for raising technological and educational standards. It may constitute the only practical path for them to engage meaningfully in international trade and to promote their long-term economic development.

There is no very satisfactory way of measuring the broader economic importance of mining to a mineral host country. However, the International Council on Mining and Metals (ICMM) has sought to fill the gap by creating a 'mineral contribution index' using data for metallic minerals, metals and coal. The index is based on three variables: minerals as a share of total exports, the recent change in this share of exports and the value of mineral production expressed as a percentage of GDP. (It should be pointed out that the value of mineral production is not strictly comparable with GDP in as far as the first is a measure of income while the second is a measure of value added.) Using these indicators, the ICMM has computed an overall index score and used this as the basis for the ranking in 2012 shown in Table 7.2.

Table 7.2 Ranking of countries according to economic contribution of mining

Rank	Country	Minerals as a % of exports 2012	Change in minerals export share 2007–2012	Production value as % of GDP 2012
1.	Mauritania	63	11	58
2.	Guyana	59	21	22
3.	DR Congo	82	12	18
4.	Suriname	76	19	16
5.	Guinea	60	9	35
6.	Bukina Faso	46	45	16
7.	Botswana	92	6	29
8.	Lesotho	45	38	15
9.	Liberia	24	21	29
10.	Australia	57	9	10
11.	Mongolia	83	2	53
12.	Papua New Guinea	51	4	26
13.	Mozambique	48	7	5
14.	Central African Rep	44	11	3
15.	Namibia	53	2	12
16.	Bolivia	30	4	11
17.	Tanzania	35	4	9
18.	Uzbekistan	19	4	11
19.	Indonesia	20	3	6
20.	Senegal	13	7	4

Source: ICMM (2014). A few countries have been omitted on grounds of size or because of lack of supporting data.

As might be expected, the list is dominated by emerging market countries in Africa, Latin America and Asia. In fact, Australia is the only advanced economy country to appear in the list. Clearly, what happens in mineral markets and in the mining sector matters greatly to these countries. Rising commodity prices in the boom years brought immediate financial benefits while at the same time serving to increase the number of countries seeking to enjoy such benefits.

Over the period 2004–2013, when the world economy achieved an annual average growth of 3.6 per cent and the advanced economies grew at an average of 1.4 per cent, most mineral-producing emerging economies performed significantly better. In Latin America, Chile grew at 4.6 per cent and Peru at 6.8 per cent. In Asia, Mongolia grew at 9.1 per cent while in Oceania Papua New Guinea grew at 6.4 per cent. In Africa, Zambia grew at 6.4 per cent, DR Congo at 6.2 per cent and Ghana at

7.8 per cent. A number of these countries, including Peru, Papua New Guinea and Zambia, managed to continue to grow through the years of the global financial crisis (GFC). Australia was, with the Republic of Korea, the only advanced economy country not to see its economy shrink in 2009.[7]

The impact of large-scale mining activities on small economies is, however, potentially double edged. Some countries such as Chile and Botswana appear to have managed the benefits of their mining industries rather well, using the wealth generated in the sector to leverage broader economic and social development and promote economic diversification. In addition to growing its mining sector, Chile has developed successful export industries in forest products, fish products, wine and tourism. However, this is far from always being the case. As discussed in Chapter 1, there is a significant body of evidence demonstrating that, for some countries, mining investment is as likely to damage as to benefit its economy. Those arguing this case tend to talk not about mining's 'contribution to' the economy but rather the economy's 'dependence on' minerals, a notion having much more negative connotations. Being dependent on minerals implies that that these economies are vulnerable to mineral price volatility, to the distortions (and opportunities for corruption) that arise when capital-intensive projects are foisted onto immature economies and to decisions made elsewhere by foreign investors.

Although such negative impacts from mining are normally associated with emerging economies, it is not their exclusive preserve. Paul Cleary, in his book quoted above, castigated the Australian authorities for allowing the proceeds of high commodity prices during the boom years to fuel a consumer and property boom to the cost of future generations of Australians. No attempt was made to neutralise the effect of rising commodity prices on the Australian dollar with the result that the manufacturing sector suffered even as the mining sector soared. Manufacturing output in 2014 was below where it was in 2002.[8] General Motors, Ford and Toyota all declared their intention to close their Australian vehicle assembly plants.

Since the idea of the resource curse gained currency in the 1990s, and particularly since the publication of the influential Sachs-Warner paper of 1995 referred to in Chapter 1, there had been plenty of time for researchers to look into the mechanisms of the resource curse on a country-by-country basis and to ask the question how resource wealth could be better managed. The MMSD report of 2002 acknowledged this as a critically important issue. It noted, 'Minerals have the potential to contribute to poverty alleviation and broader economic development at a national level', but for this to be achieved '...appropriate frameworks for the creation and management of mineral wealth must be put in place'.[9] In other words, the outcomes were largely a matter of governance. They could not be relied upon to flow automatically.

Motivated as much, and probably more, by what was going on in the oil sector than what was going on in mining, the development community was making a real effort to find out what exactly was behind the resource curse and to isolate the key elements of policies for successful national resource management. This was commonly executed through a series of country case studies designed to draw out the lessons of what appeared to work and what didn't. In 2007, Jeffrey Sachs returned to the subject with a co-authored book entitled *Escaping the Resource Curse*.[10] In 2011, Paul Collier and Anthony Venables published *Plundered Nations? Successes and Failures in Natural Resource Extraction*.[11] In 2013, the consulting firm McKinsey produced a study entitled *Reversing the Curse: Maximizing the Potential of Resource-Driven Economies*.[12]

These studies were rich with insights, and there were many recurrent themes. All the studies accepted that resource development could be a force for good, but for this to be the case certain conditions had to be met. First, mineral host countries needed effective institutions to regulate the mining sector, good laws and stable taxation, consistently and honestly administered. Second, fiscal revenues from the mining sector needed to be directed towards long-term investment (e.g., infrastructure) and to provide benefits across the population as whole so that the benefits of mining were evident to all. Third, countries needed to ensure that mine development did not come at the expense of economic diversification. In cases where rising export revenues from commodities boosted exchange rates and damaged the export prospects of other sectors, governments should consider seeking to neutralise these effects using currency stabilisation schemes, schemes whereby public funds are directed offshore into 'rainy day' accounts or sovereign wealth funds.

Unquestionably, meeting such conditions was challenging for some countries. But it was hard for them to argue that the knowledge base to enable them to avoid the curse was not available. Perhaps, surprisingly, there were very few attempts among mineral-rich countries themselves to seek systematically to draw lessons from the emerging body of research on this matter or to lay out a clear rationale for their policies towards the mining sector. An exception to this was *Africa Mining Vision* (AMV).

Africa Mining Vision was published in February 2009 by the African Union after a series of consultations among member states.[13] The 'vision', which the report promoted, was the 'transparent, equitable and optimal exploitation of mineral resources to underpin broad-based sustainable growth and socio-economic development'. However, more important than this general statement of intent, the report sought to explain in detail how mining could contribute to development in Africa, why therefore it should be encouraged and what needed to be done to attract it and ensure it served development objectives. It discussed, for example, the need for ensuring the availability of good geological data, the need for improving countries' contract negotiating capacity and the quality of their administration, and the importance of developing upstream (supply), downstream (processing) and sidestream (infrastructure) linkages from mining. While emphasising the importance of ensuring that mining countries obtained a fair share of the rents from mining, the AMV acknowledged that most countries in Africa were obliged to look to foreign companies to assist with finance and technology and emphasised the importance of partnership between the state, the private sector, civil society and local communities and of 'increas[ing] private sector confidence and participation'. In conclusion, it laid out a series of practical actions needed to take the 'vision' forward and discussed the potential of 'development corridors' (where development is clustered around major transport links) as a means to overcome infrastructure challenges in developing Africa's mineral resources.

Government responses to rising prices

The Simandou iron deposit in Guinea is the richest undeveloped iron ore deposit in the world. Rights to explore the deposit were granted to Rio Tinto in 1997. Exploration confirmed the quality of the deposit, and Rio Tinto proceeded to an assessment of the feasibility of developing it into a mine. As so often in Africa, the major challenge was infrastructure. The mine required a 650 kilometre railway line, some of it through mountainous terrain, to bring the ore to the coast for shipping. Guinea's shallow coastal waters meant it also required an unusually long wharf to accommodate the large bulk carriers that would be needed to take the mined ore to market. Without some financial assistance with the infrastructure, or other users of the rail line who could share the construction costs, it was difficult to see how the project economics could be made to add up.

In 2008, impatient with the delays, the Government of Guinea under President Lansana Conté stripped Rio Tinto of two of its four licenses for the Simandou deposit. These licences were transferred to BSG Resources, a company controlled by Israeli diamond magnate Beny Steinmetz, before being on-sold in April 2010 to Brazilian iron ore miner Vale. Following President Conté's death in 2008, Guinea was subject to two years of military rule before elections in 2010 brought a new government to office. The incoming government of President Alpha Condé was committed to review all the mining agreements made under previous administrations. After extensive discussions, the new government in 2011 agreed to recognise Rio Tinto's two remaining mining concessions, in return for which the government would acquire the right to purchase an additional 20 per cent of the Simandou project (this on top of the 15 per cent free carried interest it was entitled to under the original agreement), a one-off payment of $700 million and an agreement from Rio Tinto to formally relinquish its claim to the two concessions now held by Vale. Soon after this agreement was struck, Rio Tinto announced that it was bringing the Chinese company Chinalco into the project to provide financial assistance but also, it must be assumed, to provide itself with some heavyweight political backing.

Simandou illustrated well the dilemma with which many emerging markets were grappling, namely how best to use the opportunity presented by rising commodity prices to provide maximum long-term benefit to their economies. The availability of a growing body of work intended to help guide good policy-making in mineral-rich countries was one thing. To apply the lessons of such research on the ground in the midst of the euphoria associated with upward spiralling commodity prices was quite another. In the real world, government policies have a tendency to come piecemeal in response to particular events or to public expectations. And they are as likely to be shaped by fear of being taken advantage of as they are a thoughtful assessment of long-run costs and benefits.

As commodity prices surged, the first response of most mineral host countries was to decide that they were not getting their fair share of the proceeds and that the current rates of taxation were too low. Taxation of the mineral sector can be complex, being often applied at different levels of government: local, regional and central. It takes two principal forms: royalties based on production volumes (turnover) and corporation income taxes based on profits. As of 2012, royalties were typically in the range 2–6 per cent while corporation tax ranged from 20 to 35 per cent.[14] A few countries also operated 'excess profits' taxes designed to capture windfall gains during period of very high commodity prices.

During the second half of the 2000s, it would not be an exaggeration to say that virtually every emerging market country with a significant mining industry took the opportunity to raise its tax rates for the sector. Brazil, Chile, Peru, Zambia, Ghana, Sierra Leone, Russia, Kazakhstan, Poland, China and India were among those increasing taxes. As just one example, Zambia in April 2008 increased royalties on mining from 0.6 per cent to 3 per cent and the corporate tax rate from 25 per cent to 30 per cent. At the same the time, it introduced a windfall profits tax. Royalties were further increased to 6 per cent in 2012, and in late 2014 the government announced an additional increase of 8 per cent. Nor was the tendency wholly confined to emerging market economies. The Labour Government in Australia sought in 2010 to introduce a new 'super profits' tax on the mining sector. This proved to be highly controversial and eventually cost the Labour prime minister, Kevin Rudd, his job. A watered-down version of the tax, the Minerals Resource Rent Tax (MRRT), was subsequently introduced in 2012.

The mining industry was, as noted in Chapter 4, not entirely blameless in stoking up expectations of a commodity bonanza. Its talk of a price super cycle, with all that this implied about the increased profitability of the industry, was pretty much an open invitation for governments to up their take. The gold industry subsequently determined that its historical practice of talking about production cash costs, instead of total costs (which include capital expenditure, exploration, corporate costs and cash taxes) had aggravated matters by making the gold industry appear more profitable than was actually the case, and in 2013 the World Gold Council proposed a new industry convention for reporting 'all-in costs'.[15]

While the commodity boom lasted, the industry could in general handle increases in the tax take. There was enough to go around for everyone, albeit that not all benefited equally. An analysis of the four big UK-listed mining companies, BHP Billiton, Rio Tinto, Anglo American and Xstrata, by Deutsche Bank into who benefited from the commodities boom over the period 2005–2010 showed the government take in taxes rising a cumulative 447 per cent over the period. The next largest beneficiaries were senior management and employees with cumulative increases of 323 per cent and 210 per cent, respectively. Those benefiting least were shareholders who saw a cumulative increase of 174 per cent.[16]

Road to resource nationalism

If the issue of ensuring the revenues and profits from mining were distributed more fairly was a central feature of mining countries' policy response to the commodities boom, it was far from being the only one. The determination of mineral host governments to derive the maximum benefit from the resurgence of mining was also leading to calls for the assertion of more domestic control over the mining sector and for programmes to ensure that more mine products were processed domestically as a means to garner more value added in the local economy. Collectively, these policies were referred to by those subject to the changes – although never by those applying them – as expressions of resource nationalism.

With regard to initiatives of mineral host countries to assert a firmer grip over mining activities within their borders, these showed a significant difference from the wave of resource nationalism which had taken place in the late 1960s and early 1970s when there had been widespread instances of enforced state ownership, or nationalisation.[17] Certainly, there were some instances of outright nationalisation. For example, the Bolivian Government nationalised several mining assets, in Venezuela the gold sector was taken into state ownership in 2011, while Newmont Mining had a gold venture expropriated by the Uzbek government in 2006. However, the approach this time around was more nuanced. It was focused more on the state taking minority stakes in mining companies and on indigenisation, that is, ownership by nationals. Possibly, this was because of the demonstrable failure of the earlier nationalisations, or perhaps it reflected a view that companies which were locally owned, or had a significant local ownership, were more amenable to government influence than were foreign companies and did not therefore need to be taken into state ownership. Such a policy could give the state greater control without sacrificing the incentive structures and financial disciplines of the private sector. The target of these measures, in short, was not so much capitalism as foreigners.

As was the case with tax increases, the list of countries implementing policies to assert greater local control over their mining industries was long. In 2012, Zimbabwe began implementing a law that was originally passed in 2008 which required that all mines be 51 per cent indigenously owned. The government also determined that no new licences should be awarded to foreign companies. Mongolia in 2012 imposed a 49 per cent cap on foreign ownership of strategic mines. In the same year, Indonesia announced plans that all mines operating under Mining Business Permits (the vast majority of mining companies in Indonesia) must be 20 per cent locally owned after five years and 51 per cent after ten years. In Burkina Faso, Guinea, Sierra Leone and DR Congo, legislation was passed giving the state the automatic right to a free carried interest in any new mining project as well as the right to purchase a larger share if it chose to do so later. For Guinea and Sierra Leone, the free carry was to be 15 per cent and for Burkina Faso 10 per cent. For DR Congo, it was 5 per cent but rising to 35 per cent. In South Africa, although the government resisted demands from within the African National Congress (ANC) for wholesale nationalisation of the mining sector, its programme for black economic empowerment (BEE) was designed to secure the gradual indigenisation of major South African businesses. Companies were required to have a 26 per cent black shareholding by 2014.

Here again, this was not a tendency confined to emerging market economies. The Foreign Investment Review Board (FIRB) in Australia gives the government the right to block foreign ownership of local assets where it is deemed not to be in the national interest. Although these powers are used sparingly, they were used to block the acquisition of the Prominent Hill copper mine and rare earth producer Lynas Corporation by Chinese companies. The Canadian government, as noted earlier, in 2010 blocked the attempt of BHP Billiton to take over the Potash Corporation of Saskatchewan (PCS). To show that this cuts both ways, PCS was itself blocked by the Israeli government from taking over Israeli potash producer ICL in 2013.

Aside from the direct assertion of control over mining assets, there are many other more indirect means by which governments are able to favour local investment interests and disadvantage foreign ones, even, sometimes, while paying lip service to encouraging inward FDI. They can, for example, impose strict 'use it or lose it' provisions on the licenses they issue as was the case with Rio Tinto's Simandou project in Guinea. Incoming governments in Guinea, DR Congo, Liberia, Sierra Leone, Ghana, Kenya, Madagascar and the Dominican Republic have all used the wholesale review of previously awarded licenses as a means to extract new concessions from foreign miners. Russia ostensibly encourages inward investment in mining, but uncertainties over its mining laws and their application, coupled with more general provisions applying to all foreign investors, mean that in practice it remains a difficult place for foreign miners to operate. A number of minerals such as uranium, diamonds, rare earths, nickel, cobalt, platinum and tantalum are off-limits to foreigners for strategic reasons while special restrictions apply to the mining of large deposits of other commodities such as copper and gold. China similarly pays lip service to the importance of inward investment in mining, but administrative obstacles and ferocious local competition have restricted the interest of overseas investors in the sector. Brazil provides financial incentives to encourage its companies to explore and invest at home rather than to head overseas.

The other major plank of mineral host countries' attempts to extract greater value from mining feature schemes to encourage the local processing – 'beneficiation' as it is called in the industry – of domestically mined ores as a means to secure additional value added for the home economy. For many emerging economies, the logic of this is self-evident. The export of unprocessed raw materials is perceived as a hangover from the colonial era. If a country has mineral resources, it should also have the economic benefit of processing these through to usable forms rather than see that benefit enjoyed elsewhere. In an ideal world, it will use the final products in its own domestic manufacturing industries too. The case for domestic processing in mineral producing countries has long been a mantra of the development community and supported by agencies such as UNCTAD. One of the objectives of the *Africa Mining Vision* was to 'promote local beneficiation and value addition of minerals' and make 'use of the locational advantage of producing crude resources to establish resource-processing industries (beneficiation) that could then provide the feedstock for manufacturing and industrialisation'.[18]

The economic logic for domestic processing is, in fact, far from clear-cut. The factors making for competitive advantage in metal production are fundamentally different from those for mining. For mining, competitive advantage is all about resource quality and infrastructure. For processing, competitive advantage is more about power availability, cheap capital and technology. Cost curves for metal processors are typically flatter than those for miners and margins narrower, meaning that to make money plant has to be run at close to its limit. It may well be that value can be added by processing, but the key question from an economic perspective is not whether value is added but whether the activity is wealth creating. Value added is essentially an accounting artefact; it does not pay regard to the capital required to generate the value. Wealth creation is an economic concept, which requires not only that there is value added but that there is a return on capital as well. From the economist's perspective, therefore, beneficiation should be considered case by case, having regard to the objective circumstances of each project and each country.

While there may be scope to debate the rationale for their actions, it nonetheless remains the case that many mining countries have made it a policy objective to seek to develop downstream processing industries based on locally mined resources. In some cases, the requirement for a mineral to be processed locally comes as a condition of mine permitting. Among the many countries announcing plans to require some form of in-country beneficiation are South Africa, Indonesia, Brazil, DR Congo and Vietnam.[19] In 2013, DR Congo sought to impose a ban on the export of unprocessed copper and cobalt, although in the absence of sufficient smelter capacity and power in the country to process the ores domestically the implementation of this measure had to be deferred.

A popular mechanism for tilting the economic playing field in favour of local processors is to restrict exports using export taxes or quotas, thereby making the products cheaper in the domestic market than in the export market. A number of mineral-producing countries employ these measures. The World Trade Organization (WTO), in its annual report of 2010, found that the incidence of export restrictions among natural resource producers was increasing.[20] India, for example, imposes tariffs on exports of iron ore and steel scrap to ensure the availability of raw material supplies to local steel producers. Russia for years imposed export tariffs on many forms of scrap as well as on refined copper, nickel and PGMs, although these were scheduled for removal subsequent to Russia's accession to the WTO in 2012. China has made extensive use of export restrictions (tariffs and quotas) across a range of mineral products, as discussed Chapter 3. The government of South Africa in 2013 announced plans to amend the Mineral and Petroleum Resources Development Act (MPRDA) to give ministers greater direct control over mineral exports as a means to promote the domestic processing of minerals.

One of the more high-profile examples of the implementation of such policies is provided by Indonesia, a country which has seen rapid growth in its production and export of minerals such as coal, copper, nickel and bauxite. In the first half of 2012, the Indonesian government announced a series of measures aimed at asserting greater direct control over its mining sector and retaining more of the benefits of its production of minerals in the local economy. These measures included the immediate imposition of a 20 per cent tax on the export of unprocessed minerals and a stipulation that all exports of unprocessed minerals would be banned from 2014. In the meantime, only those companies committing to beneficiate locally would be permitted to export at all. Despite loud protests from mining companies and some scepticism that the timetable would be adhered to, in January 2014 the ban on the export of unprocessed ores from Indonesia came into force. It was accepted that the loss of export revenues and mining jobs would have a negative effect on the economy near term, but the government believed that these would be outweighed by the longer-term development benefits to the Indonesian economy.

Structural basis for resource nationalism

These instances of resource nationalism paint a compelling picture of a widespread assertion of sovereign control over mineral resources and of an attempt by mineral-rich countries to extract greater value from their domestic resources. Unquestionably, the commodity boom was a major factor driving this. However, it was far from being the sole cause. If it were, then it might be expected that the push for resource nationalism would fade as commodity prices weakened. While it may well be that some countries will adjust their policies for the mining sector in response to a weaker price environment, there has been no general retreat from resource nationalism.

The resurgence of resource nationalism in mining has tapped into other, broader tendencies in the global economy. It would also appear that there are some longer-term, structural factors at work.

The GFC brought with it a general disenchantment with economic globalisation and a growing distrust of the multilateral institutions designed to promote it. In the wake of the crisis, many countries chose to prioritise their own immediate economic interests. The rate of growth in international trade slowed to well below the 7 per cent a year it had averaged over the previous 30 years.[21] Although not taking on the extreme form that it did in the 1930s, there was nevertheless clear evidence of a return to more protectionist thinking. This was apparent in discriminatory trade policies, in restrictions on foreign lending by banks, in the preferment of local companies over foreign ones and in attitudes towards immigration.[22] There was a backlash against FDI, both among exporters of capital and among importers. While the effects of this were felt across all sectors, it had particular relevance for the resources sector. In blocking the sale of a Canadian oil-sands company to a Chinese state-owned company, Canadian premier, Stephen Harper, said, 'When we say that Canada is open for business, we do not mean that Canada is for sale to foreign governments'.[23] The president of the African Development Bank (AfDB), Donald Kaberuka, was even more forthright in his views about what he saw as the consequences of international investment, claiming in the middle of 2013, 'The reality is, Africa is being ripped off big time.'[24]

Resource nationalism was also tapping into a wider tendency for the state to play a greater role in the orchestration of economic life, especially in emerging economies. This was not quite the tendency towards nationalisation, often undertaken for ideological reasons, which had been evident in the 1970s. It was about a pragmatic desire to have the state become more involved in directing and coordinating the activities of industry so as to ensure that they pulled together to serve the interests of the national economy. It was tendency often described as 'state capitalism'.[25]

The Economist in a special supplement, in 2012, talked of state capitalism as an attempt 'to meld the powers of the state with the power of capitalism. It depends on government to pick winners and promote economic growth. But it also uses capitalist tools such as listing state-owned companies on the stockmarket and embracing globalisation'. Noting that state capitalism could claim China, the world's most successful big country, for its camp, it commented, 'They think they have redesigned capitalism to make it work better, and a growing number of emerging world leaders agree with them.'[26]

In addition to these two broader geopolitical tendencies which were underpinning resource nationalism there appeared to be a third, structural factor.

The structural foundation of resource nationalism is to be found in the fact that recent years have brought with them increased polarisation between mineral-consuming and mineral-producing countries.[27] Historically, the major producers of minerals tended also to be the largest consumers of minerals, furnishing a strong incentive for investment in minerals to supply low-cost raw materials to the domestic economy. The industrialisation of Europe and North America was based largely on mineral raw materials supplied locally.

However, over the past 50 years, producing and consuming countries have increasingly been driven into separate camps with distinct and differing interests. This is partly the result of the industrialisation of Asian countries which are possessed of limited domestic mineral resources but it is also reflects the fact that the focus of production growth has been on areas with relatively low mineral consumption, notably Australasia, Latin America, Africa and Asia, excluding Japan and China.

As shown in Figure 7.2, in 1960, the world's major consumers of minerals – defined for present purposes as North America, Western Europe, Japan, China and the Former Soviet Union – supplied 82 per

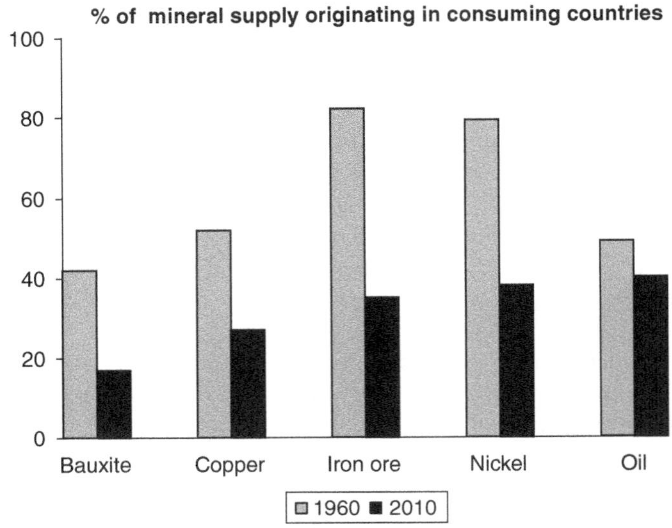

Figure 7.2 Polarisation of global mineral supply
Source: Humphreys (2013).

cent of their own iron ore, 78 per cent of their nickel and 52 per cent of their copper. By 2010, these proportions had fallen to 35 per cent, 38 per cent and 27 per cent, respectively; that is, they had effectively halved. Given where the undeveloped resources of the world are found, this is a trend that is likely to continue.[28]

These developments have had two important and related consequences. The first is that they have created a lot more countries which are economically dependent on resources. In their study, *Reversing the Curse,* the McKinsey Global Institute claimed that there were 81 resource-driven economies in 2011, up from 58 in 1995.

The second consequence, which follows from the first, is that there are more countries than ever looking to mineral resources to provide a springboard for their development. For these countries, mining is no longer about the provision of cheap raw materials for domestic industries but about the fulfilment of their national economic potential. In some cases, mining may be the only viable basis for economic development available to them.

This change in industry structure thus implies a fundamental change in industry narrative. The matter at issue now is not what the mine will produce or how profitable it will be, but the contribution it can make to national economic development. Within the context of this narrative, governments will almost inevitably want to assert a firm guiding hand over the mining industry and plan the manner of its interactions with the rest of the domestic economy. A consequence of this shift of perspective is to make the sector inherently more political. In effect, it is moving closer to the model of oil, which has been living with a polarised structure of supply (see Figure 7.2) and a condition of acute politicisation rather longer than has the hard minerals sector.

Corporate response

The cumulative effect of this tendency for countries to seek more from their minerals – and its expression in resource nationalism – clearly constitutes a major challenge for the international mining industry on several different levels. Rules restricting the ownership of assets by foreigners or offering preferment to local companies may not in themselves kill off projects, but where the requirement for a local shareholding involves a significant free carry, it raises the hurdle rate which a project must achieve if it is to be profitable for other investors. High levels of taxation present a similar problem. While corporate income taxes take a share of a company's profits and will ride up and down with commodity prices, royalties are a direct addition to cash costs and will impact directly on margins and on the ore grades that a company can profitably mine. 'Use it or lose it' provisions in mining agreements, while understandable for countries impatient for development, ignore the inherently cyclical nature of the industry and mean companies lose flexibility in the preparation of projects.

If onerous project conditions can be a problem, in a capital-intensive industry such as mining uncertainty over those conditions can be just as damaging. Uncertainties over the tax rates to be applied to a project make it impossible to evaluate properly. Uncertainties about the robustness of agreements made, and whether they could become subject to 'review' a year or two after being struck, will leave companies and their bankers reluctant to commit to large-scale expenditure. Foreign investors have most negotiating leverage before they commit to a project. This leverage diminishes as a project is built and moves to production leaving the investors highly exposed financially to changes in the rules at a point where there is little they can do about it, a problem known in the oil industry as the 'obsolescing bargain'.[29]

The combination of these pressures together with a weaker price environment after 2011 inevitably led mining companies to back away from some of their more adventurous projects in higher-risk countries. BHP Billiton, Rio Tinto, Vale and Anglo American all closed down their exploration programmes in DR Congo between 2011 and 2013.[30] Kinross Gold withdrew from a $1 billion gold-silver project in Ecuador citing high taxes as the reason.[31] Antofagasta and Barrick Gold abandoned the Reko Diq copper-gold project in Balochistan, Pakistan, in 2013 after years of wrangling about the terms of development.[32] Vale put on hold its work on the two concessions it had acquired for the Simandou project before its rights were stripped from it by the Guinean government in 2014. The development of the second phase (the underground mine) at Rio Tinto's Oyu Tolgoi stalled as the company and the government struggled to reach agreement on the terms for its development.

If the first wave of the commodities boom up to 2008 prompted the revival of resource nationalism, it was reinforced during the second wave starting late 2009. An annual survey of business risks in the mining industry by accountancy firm Ernst & Young showed that between 2008 and 2011 resource nationalism rose in the rankings of the major risks confronting the sector from eighth place to first, a position it retained in 2012. It remained a major concern in 2013 and 2014, albeit that more immediate financial pressures and a lesser emphasis on growth had by then pushed it off top slot.[33]

The mining industry was not wholly unprepared for the growing politicisation of the sector. But the policies it had come to in adversity during the 'barren years' increased massively in importance with the greater confidence and assertiveness of host countries during the boom years. As described in Chapter 1, it had already become clear by the end of the 1990s that the industry suffered from deeply negative public perceptions and needed to improve its performance. Some aspects of this were competitive. Companies which showed themselves sensitive to the issues of mineral host countries and which had solid track records in dealing with host country concerns hoped to benefit from this by positioning themselves as 'developers of choice'. But there were also important non-competitive issues relating to the perception of industry as a whole. It was doubly hard for a company to establish a positive reputation for itself in an industry that was perceived negatively. And there were many non-competitive areas where those concerned about perceptions of the industry could productively collaborate, for example on environmental and product research.

Admittedly, the attempt to tackle these matters at a cross-industry level was made difficult by the broad span of the industry. Although the agenda for change was driven by the large publicly listed mining companies, the public perception of the industry was shaped as much by the behaviour of small-scale unregulated artisanal miners, by miners operating in conflict zones, by junior mining companies with limited budgets and a predisposition to cut corners and by state companies which lacked adequate checks on their activities.

Nonetheless, during, and in the wake of, the Global Mining Initiative of 2000–2002, enormous efforts were made by the major mining companies to raise their standards with respect to all aspects of their business activity. For an industry dealing in depleting assets, the ability to position itself within a sustainable development framework was critical and this, accordingly, was the framework adopted for much of this work. The reports and websites of the companies concerned, as well as those of the CEO-led industry organisation, the ICMM, provided extensive descriptions of sustainable development policies and programmes and testified to the enormous efforts that the industry had made to raise its game.

Among the many focuses of this work were programmes to engage with communities in the vicinity of mining operations and to encourage local suppliers. Special efforts were directed towards the rights and needs of First Nation and indigenous peoples. There were also programmes devoted to environmental concerns, covering water use, emissions to the air, biodiversity, closure of mines and the rehabilitation of mine sites. Other programmes addressed issues of health and safety and product use, reuse and disposal. Overarching programmes sought to improve the public reporting on all these matters and the transparency of company dealings with, and payments to, governments. Many mining companies adopted the standards of the widely recognised Global Reporting Initiative (GRI) and, from 2010, supplementary standards devised specifically for the mining industry, drawn up by GRI and ICMM. The pursuit of these policies, the companies believed, was necessary for them to earn a 'social licence of operate' in mineral host countries. Arguably, they were going further than any industry in history had been required to go in thinking through, and addressing concerns about, the consequences of its activities and investments.

Building partnerships

A commonly recurring theme in much of this work was one of 'partnership', specifically partnership between the companies and the peoples in the communities, regions and countries in which they operated.

This was also a key plank in the companies' attempts to address the challenges arising from resource nationalism.

The notion of partnership in this context rested on the principle that, in return for access to a country's resources, a company would partner in its economic development by providing skills, organisation and financial resources. This idea went well beyond the idea that it was simply making an investment, creating jobs and making a contribution to macroeconomic growth (GDP) and exports. It suggested an active and positive engagement in the broader development aspirations of the host country. It was something that needed to be based on a clear understanding of the microeconomic possibilities of mining ventures and their impacts on upstream and downstream industries. But it also needed to consider how mining investments could be used, through providing shared infrastructure such as ports, roads, rail and power plants, to leverage development in other sectors of the economy, sectors such as agriculture, forestry, fishing, tourism and manufacturing, which could continue to be productive after the mine closed down.

The need to shift the focus to the broader development potential of mining arose from several sources. First, the scale of some mining projects relative to the size of the local economy had grown to oil industry dimensions. Rio Tinto's Oyu Tolgoi mine in Mongolia is anticipated, at full production, to increase the GDP of Mongolia by a third.[34] Its Simandou project, should that proceed, is estimated to double the size of the Guinean economy and (excluding employment taxes and other fees) double direct payments to the Government of Guinea.[35] The impacts from such projects cannot be left to chance; they have to be planned for. Second, there was the need to move the discussion of the industry's economic contribution beyond the matters of tax take and community expenditures to the broader issues of wages and payments to suppliers. Third, there was a need to develop policies to combat the effects of the resource curse. For these reasons, large mining companies had begun to employ regional and development economists to enable them better to understand their projects' potential economic impacts and to provide guidance on their management.

Just as the development community had been asking the question how to overcome the challenges of the resource curse, the mining industry had itself been seeking to contribute to the analysis of the problem and the identification of solutions. In 2004, the industry body, the ICMM, launched a programme – the Resource Endowment Initiative – to probe the issue of how best to manage mineral resource wealth through a series of case studies on individual countries, to determine what worked and what didn't. Countries investigated included Brazil, Chile, DR Congo, Ecuador, Ghana, Guinea, Laos, Peru, Romania and Tanzania. On the basis of this research, the ICMM issued a 'toolkit' in 2006 (revised in 2011) for mining companies, laying out guidance on how to maximise the benefits and minimise the negative effects, of mining developments.[36] It covered such issues as the organisation of community programmes, assistance with regional economic management, the development of social programmes (such as those relating to health and education) and how best to engage local businesses.

A position paper published by the ICMM is June 2012 revealed a subtle but important shift in emphasis away from the industry's concerns with the sustainability of its activities towards a broader development agenda. In the words of the report, 'The focus is not now on how mining can be sustainable... but on how mining, minerals and metals can contribute to sustainable development. This is a conceptual shift away from a singular analysis and mitigation of impacts to a more comprehensive analysis that looks at the wider contribution of the industry and its products. The focus on contribution is a tougher but fairer approach. This is so because it demands a demonstration of positive results and not just mitigation of negative impacts.'[37]

This issue of the economic contribution of the mining sector was also taken up by the World Gold Council (WGC). In a study first published in 2013, the WGC provided a detailed breakdown of spending by its members on payments to government, purchases from other businesses, wages and salaries and community investments, as well as payments made to providers of capital.[38] These are shown for the years 2009, 2012 and 2013 in Table 7.3. Naturally, there is some variation in these numbers year to year subject to fluctuations in the gold price (which fell sharply in 2013) and there is some small variation in the company coverage. With this caveat, direct payments to government comprised 15 per cent of total company disbursements in 2012 and 10 per cent in 2013. In both years, by far the biggest disbursements, accounting for around 80 per cent of the total, were payments to other businesses and to wages and salaries. The economic impact of these latter disbursements – the vast majority of which are made in-country – is of the essence in the assessment of mining's economic contribution.

The idea of the mining industry positioning itself within a development framework was not without its critics. Some within the industry were deeply, and understandably, attached to the idea of operating within a strict law-based system. This system laid out clearly the duties and obligations of all parties to an agreement and backed them up

Table 7.3 Expenditure by members of the World Gold Council, in $ million

	2009	2012	2013
Royalty payments to government	542	1,389	811
Income tax payments to government	3,826	5,494	2,509
Other payments to government	961	1,575	1,330
Payments to businesses	19,690	35,157	32,440
Wages and salaries	4,098	8,290	6,054
Community investments	121	285	340
Payments to providers of capital	814	3,404	3,813
Total expenditure	**30,051**	**55,594**	**47,297**

Source: WGC (2013b, 2014).

with the force of law. It provided for a clear demarcation between the role of the company and the role of government and thus restricted the scope for confusion and fraud. It permitted a company to retain a respectful distance from political discourse within the country. Against this background, some in the industry felt it inappropriate that mining companies should seek to position themselves as de facto development agencies.

The problem was that for a number of countries in which international mining companies wished to operate, such a strict law-based approach was simply not on offer. In these countries, mining agreements had an inherently political character. While they permitted the development of more organic collaborative relationships between the parties than were typically available under a strict rules-based approach, they lacked the certainties and protections that a law-based system provided, or at least should provide. This may not be an issue for a Chinese company with the support of the state, but it was very much an issue for a publicly listed company without any visible political backing.

The shift in emphasis also pointed to the need for a different mindset on the part of the company and, to a degree, a different type of leadership. Rather than rely on the government as the sole interlocutor for projects, companies, to become a fully fledged partner, needed to invest considerable time and effort in seeking to understand the politics, culture and history of the countries in which they were hoping to operate and to engage in broad-based consultation and trust-building across a wide range of interest groups. To establish their credentials as a partner, they would, in short, have to convince people that they were serious about their involvement in the country and committed to it long term.

Only by this means were they likely to build within the country an understanding about their project and develop the grass-roots support they needed. Their experiences elsewhere would help only to a limited degree. Every country tends to have its own demands and expectations.

With the best of intentions, companies on their own cannot make partnerships work. Effective partnerships require good faith on both sides over a sustained period. They need the involvement of governments who, in determining what is a 'fair' division of rents, have a clear understanding of the value of the capital and skills that companies bring and some appreciation of the fact that shareholders provide money to companies in the expectation that they will get a return. Companies cannot commit to major investments without the support of their shareholders, and investors have many options other than mining for their money. They need governments who provide policy continuity and which take a strategic rather than opportunist view of the value of the industry; who recognise that the economic benefits of mining come not just from the taxes companies pay but from the entirety of their activities, the wages they pay, the supplies they buy and the skills they impart; and who are focused on the growth of the pie not just the division of the pie. And they need governments who think long term and who are not going to be distracted by the offer of short-term deals from private intermediaries operating out of public view, which may enrich the country's elites but which do nothing for the population as a whole.

Unfortunately, mining can too easily become a political football in countries' domestic politics. A good illustration of this was provided by the presidential election in Mongolia in June 2013 when a large part of the political debate leading up to the election was focused on which party was going to be toughest on the mining sector. Other examples are provided by South Africa where debates about the industry have become a proxy for divisions within the ANC about economic and political priorities[39] and by Peru where the matter of awarding exploration and mining licences has become inextricably bound up in a debate about the rights of indigenous peoples to block developments deemed otherwise to be in the national interest. Elsewhere, mineral developments have become the focus of disputes over the distribution of benefits between central governments and the regions in which the developments occur.

In the final analysis, it may be that there are obstacles to establishing good and lasting relationships with mineral host countries that companies are simply unable to overcome. No one seriously questions the right of countries to choose how they develop their resources, who they engage with in this development or indeed whether the resources are developed at all. It is a sovereign right recognised by the UN in a series of resolutions dating back to the 1950s. Where trust between host countries and mining companies breaks down; where companies are simply not able to operate to standards acceptable to them; where the terms of an agreement are unilaterally changed in ways that render projects unviable; where governments are too weak to hold the ring between competing sectional or regional interests; and where simply being foreign is a problem, there is not too much a company can do. They may have little choice, but, as they did in the early 1970s and 1980s, to retreat back to countries where inward investment is welcome and the rule of law prevails, even if these places do not offer the best possible resources. That a deposit is the best in the world counts for little and benefits no one if it cannot be profitably developed.

8
Consumer Concerns

Minerals turn critical

> Automobiles manufactured at the start of the twentieth century were composed of about five materials: wood, rubber, steel, glass, and brass. Today, a typical automobile may contain up to 39 different nonfuel minerals in various components, in addition to rubber, plastics, and other organically based materials.... In the 1980s, computer chips were made with a palette of twelve minerals or their elemental components. A decade later, 16 elements were employed. Today, as many as 60 different minerals (or their constituent elements) may be used in fabricating the high-speed, high-capacity integrated circuits that are crucial to this technology.[1]

Modern economies are highly dependent on regular supplies of a wide range of mineral products. The rapid growth in China's demand for minerals and the escalation of mineral prices to which this gave rise, while it may have been a boon to mining countries and companies, was a source of growing concern among the traditional mineral-consuming regions, the United States, Western Europe and Japan. These concerns were aggravated by growing evidence of resource nationalism and the increased role of the state in mineral-producing countries.

The concerns of the traditional mineral-consuming countries arose at a number of levels. On one level, it was simply a response to complaints from local manufacturing companies that higher prices of mineral commodities were raising their costs and squeezing their profit margins. On another, there was a concern that shortages of certain commodities – or even their non-availability – represented a long-term strategic threat to a country's basic industrial and technological capabilities. Finally,

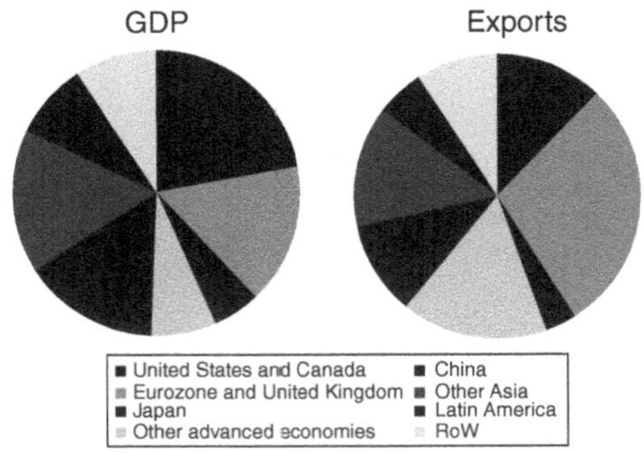

Figure 8.1 Composition of global GDP and exports of goods and services, 2013
Source: IMF (2014).

there was a concern that, given the uses of some of the minerals under threat, shortages might compromise national security. Minerals deemed to have particular strategic significance, whether economically or militarily, were often referred to as 'critical'.

It may have been the case, as discussed in Chapter 2, that the economic dominance of the advanced economies was gradually diminishing, but these countries still represented almost half of global GDP, 60 per cent of global exports (Figure 8.1) and an even higher proportion of global final demand. They were also home to many of the world's largest and most advanced technology companies, responsible for designing and developing the sophisticated electronic gadgetry – the phones, tablets and computers – on which a modern consumer economy depends. Their supplies of mineral raw materials, therefore, remained a matter of considerable importance.

This chapter examines the condition of the mining sector within the large traditional mineral-consuming countries and some of the factors shaping the policy responses of these countries to the threat of mineral shortages. It then looks at the policy measures adopted by the European Union (EU) and the United States, and in particular at the measures applied in relation to rare earths. The final section offers some reflections on the coherence and effectiveness of the policy responses described.

Shrinking mine supply

If the traditional mineral-consuming countries were experiencing growing concerns about their supplies of mineral raw materials, it was partly because their ability to meet their requirement for these materials from domestic sources was diminishing. This had not been a matter of much concern through most of the 1980s and 1990s while minerals of most sorts were cheap and abundant, but in the 2000s it became a matter of very great concern indeed.

The reason these countries were less able to supply their own raw material needs is not hard to discern. As discussed in Chapter 7, the majority of the world's mineral resources lie outside the advanced economies and mineral production had been migrating to places where it could most cheaply and efficiently be undertaken. This is how international trade is supposed to work. However, a strong case can be made that the change was also the result of complacency with regard to mineral supplies and a neglect of the mining sector in the advanced economies.

As discussed in Chapter 1, a common belief seemed to have taken hold that modern economies were 'weightless' economies, which did not need raw materials in the same way that traditional industrial economies needed them. Public policy had come to reflect this belief. Encouragement of resource development was simply not seen as a priority, and, in the face of growing competition from other possible land uses, permits to open mines were becoming ever harder to obtain. In 1996, the year of his re-election, President Clinton had courted popularity by making a very public demonstration of his administration's blocking of a proposed gold mine located near, but not in, Yellowstone National Park, declaiming, 'Yellowstone is more precious than gold'.[2] As one long time observer of the industry has put it, '...it would not be going too far to say that most countries in the West seem to have gone out of their way to discourage mining through the complex web of the government approvals process that erects countless environmental, social, and land-use barriers.' [3] The focus of industrial policy in these countries was almost entirely on downstream, high-tech activities and on services. Mining was not something governments appeared to want to support or be seen to be supporting.

Accordingly, the institutional capital of the mining sector in the United States and Europe was allowed gradually to erode. As mines closed, so too did mining support services and mining schools. There were progressively fewer and fewer people in government with a knowledge of, or interest in, mining. The US Bureau of Mines (USBM) was closed in 1995.[4] Although some of its functions were transferred to the US Geological Survey, many of them were allowed to die. The USBM's closure was a clear symbol of mining's diminished status in US government thinking. This was an awkward, unpopular industry from which governments were happy to distance themselves. The declining global influence of US and continental European mining companies and the decline of the US stock market as a source of finance for the mining industry were viewed with apparent indifference. Even in the United Kingdom, which had seen significant economic benefits arising from its increased role as a centre for mining companies and for mining finance – which in turn had generated business for a whole raft of mining-related businesses such as mining law, accountancy, consultancy and trade press – there was indifference to the sector in government.

What went for mining went also for mineral processing technologies. With a diminished interest in mining, less and less were the governments of the United States and of European countries prepared to support research into the processing of mined products. The technology for the separation of the rare earth elements terbium and europium was developed in UK government laboratories (Warren Springs) before eventually finding its way via the companies Rhone Poulenc and Molycorp to China.[5] In the 1990s, the United States was the dominant producer of rare earth magnets, based on technologies pioneered by the Pentagon. With the decline of rare earths mining in the United States, the leading producer of rare earth permanent magnets, Magnequench, a General Motors subsidiary, sold its technology and associated manufacturing operations to a Chinese group, which in 2002 dismantled the plant and transported it wholesale back to China. Although the US government had the power under the Exon-Florio Amendment to the US Defense Production Act to prohibit the sale on national security grounds, in a decision which it may subsequently have regretted, it chose not to do so.[6] In the same year, the Rare-earth Information Center (RIC) at the Iowa State University, which had provided scientific and technical information concerning rare earths to industry and government over 36 years, was forced to close its doors as a result of the loss of financial support.[7]

The data show clearly the decline in mining in the traditional mineral-consuming countries. Although the revival of mineral markets in the mid-2000s certainly breathed life back into the mining sector in certain areas – the countries of Scandinavia are a case in point – the inertia of previous actions and policies meant that there was no reversal of trends during the boom period. As shown in Table 8.1, production of many economically important minerals in these areas has declined or been stagnant since 1990. Over the same period, their collective share of global production halved or more than halved. With Europe and Japan already relatively small producers as a result of earlier mining closures,

Table 8.1 Mine production in traditional mineral-consuming countries, 1990 and 2013

	1990	% of world	2013	% of world
Copper (000 tonnes)				
W Europe	401	4.5	304	1.6
United States	1587	17.7	1255	6.9
Japan	13	0.1	0	0.0
Subtotal	2001	22.3	1559	8.5
World	8957	100.0	18294	100.0
Gold (tonnes)				
W Europe	36	1.7	17	0.5
United States	294	13.8	229	7.6
Japan	7	0.3	6	0.2
Subtotal	337	15.8	252	8.3
World	2132	100.0	3022	100.0
Iron ore (million tonnes)				
W Europe	36	4.1	31	1.6
United States	56	6.3	51	2.6
Japan	0	0.0	0	0.0
Subtotal	93	10.4	82	4.2
World	888	100.0	1930	100.0
Zinc (000 tonnes)				
W Europe	881	12.7	652	4.8
United States	543	7.8	788	5.7
Japan	127	1.8	0	0.0
Subtotal	1551	22.3	1440	10.5
World	6965	100.0	13720	100.0

Source: GFMS, RMD (SNL) UNCTAD, WBMS.

much of this decline was accounted for by the United States. Given that these countries in 2013 collectively accounted for around a third of global demand for minerals, this implied that they were becoming increasingly dependent for their supplies on imports.

Policy challenges

The growing import dependence of the traditional mineral-consuming countries clearly posed a challenge for these countries at a time when global mineral supplies were coming under pressure. A policy response was called for.

A problem faced by these countries in fashioning a coherent policy response was one of time lags. Generally, they were committed to free trade and to the rules-based system of the World Trade Organization (WTO) which supported it. In principle, so long as minerals were traded freely and fairly, this was viewed as the best way to satisfy all countries' mineral requirements. However, this stance was not without its problems. The mining industry is a slow-moving industry, and it takes a long time for supply to respond to higher prices and restore balance to markets. These lagged responses create a mismatch between the time horizons of policy-makers and what the market can deliver.

A second problem in fashioning an effective policy response to the mineral supply challenge was the ambivalent attitude of governments of mineral-consuming countries towards the mining sector. The Government of China, which, because of the country's rapid economic growth, faced a mineral supply challenge far greater than that faced by the United States, the EU and Japan, had actively intervened to promote domestic mining and to support the activities of its mining companies overseas, as was discussed in Chapter 3. These were not policies which remotely appealed to the governments of the traditional mineral consumers. Certainly, there appeared to be little appetite for an interventionist policy to encourage domestic mining. With regard to the activities of their companies' overseas, there was, again, little interest in seeking to provide support with development and infrastructure projects of the sort that China was giving its companies. In principle, the provision of funding for infrastructure to assist global mining companies with their projects was something the World Bank might have helped with. In practice, institutional divisions within the Bank, in particular the division of responsibilities for the funding of public sector projects (which come under the International Bank for Reconstruction and Development) and the funding of private sector projects (which come under the International Finance Corporation), severely restricted its ability to fulfil this role.

If governments of the traditional mineral-consuming countries were not much interested in actively promoting their mining companies in their activities overseas, there was one aspect of these companies' behaviour which was of considerable interest to them and that was their governance. Thus, for example, these governments embraced initiatives to bring greater transparency to payments made by the mining industry through devices such as the Extractive Industries Transparency Initiative (EITI), as well as through similar regional initiatives. They gave a particular focus to efforts to stamp out bribery, through legislation such as the United Kingdom's Bribery Act and the US Foreign Corrupt Practices Act (FCPA). And they encouraged and supported initiatives for ensuring tighter supply chain management in the minerals sector. Such initiatives included the Kimberley Process for ensuring the exclusion from world markets of so-called conflict diamonds (diamonds used by rebel movements to finance wars against legitimate governments) and the issuance by the Organisation for Economic Co-operation and Development (OECD) in 2011 of due diligence guidelines for companies procuring supplies of minerals from conflict-affected and high-risk areas.[8]

Response of the European Union

After China, the EU is the largest consumer of minerals in the world. The European Commission (EC) was anxious to be seen to be doing something about the issue of raw material supply in response to expressions of concern by European businesses. However, in addressing the issue of mineral supply, it was hampered by the fact that mining in the EU is still a matter of national jurisdiction. The European Commission's voice in this area is really no more than one of persuasion. So far as policy actions were concerned, it was obliged to approach the issue somewhat obliquely through policy areas where it had clearer jurisdiction, mainly environment, trade and research and development (R&D).

Following a two-year consultation with industry, the European Commission in November 2008 published what it called its 'raw materials initiative' (RMI). This took the form of a 'communication' to the European Parliament and the European Council. The RMI asserted that 'raw materials are essential for the sustainable functioning of modern societies' and that 'securing reliable and undistorted access to raw materials is increasingly becoming an important factor for the EU's competitiveness'. It went on to claim that such access to raw materials was being compromised by rapidly growing demand for mineral commodities in emerging economies, and China in particular, as well as from the protectionist policies of mineral producers. 'Increasingly, many emerging economies are pursuing industrial strategies aimed at protecting their resources base to generate advantages for their downstream industries. This is apparent in the proliferation of government measures that distort international trade in raw materials.

They include export taxes and quotas, along with subsidies, price-fixing, dual pricing systems, and restrictive investment rules.' It further noted, 'Emerging countries are also pursuing strategies towards resource-rich countries with the apparent aim of securing privileged access to raw materials'.[9]

Its proposed policy response comprised three 'pillars'. The first pillar involved seeking improved co-operation with mineral-producing countries and the removal of unfair trade practices which were inhibiting the EU's access to primary and secondary raw materials. The second was to 'foster sustainable supply of raw materials from European sources'. Because of the jurisdictional limits on what the EU could do here, this amounted to little more than encouraging better coordination of national geological surveys and promoting R&D on the extraction and processing of minerals under existing EU programmes. The third pillar was, in effect, an attempt to circumvent the supply problem by reducing the EU's consumption of primary raw materials, this to be achieved by adding emphasis to existing EU initiatives on resource efficiency and increased recycling. The RMI also contained a proposal to undertake a study 'to define critical raw materials for the EU' and 'to identify a common list of critical raw materials'.

This study was duly published in June 2010. To arrive at its list of 'critical raw materials', the EC working group used an analytical framework which had been developed during a previous supply crisis in the late 1970s and early 1980s. This involved assessing minerals according to two criteria, (a) their economic importance in use and (b) their vulnerability to supply shortage, and then plotting them out on a matrix such that the most critical minerals appeared in the top right-hand quadrant. To the extent that they were able, the working group sought to quantify these assessments. Thus, for example, the assessment of the economic importance of a mineral took into account the value added of the industrial sectors in which it was employed, while the assessment of supply vulnerability took into account the degree of country and company concentration among producers.

The final list of critical minerals which the EC working group came up with had a pronounced emphasis on minor metals used in electronic and green energy applications and on minerals coming from China. The list comprised 14 mineral commodities: antimony, beryllium, fluorspar, gallium, germanium, graphite, indium, magnesium, rare earths, tungsten, niobium, PGMs, cobalt and tantalum.[10] Significantly, the first ten of those listed were sourced by the EU substantially from China. The policy recommendations of the report essentially followed those of the

RMI, arguing for the promotion of exploration and research on minerals and mineral processing within the EU and a renewed emphasis on resource efficiency and recycling. With respect to the external dimension of the policy, it urged the EU to 'engage without reservation in consultations with third countries whose policies are causing distortions on international raw materials markets' and to 'consider the merits of pursuing dispute settlement initiatives at WTO level so as to include in such initiatives more raw materials important for the EU industry'. There was little doubt that the particular 'third country' the EU had in mind was China.

The degree of concern about these matters across the countries of the EU varied significantly. As Europe's largest manufacturer, concerns were greatest in Germany, which, accordingly, pushed hardest for substantive policy action. Under its own commodities strategy, the German government undertook to provide financial support for exploration by German firms overseas while a group of 12 German companies, including carmaker Daimler and electronics group Bosch, formed an alliance to invest in raw material projects in order to improve security of supply.[11] In 2010, the German government set up the Deutsche Rohstoffagentur (DERA – the German Mineral Resources Agency) with a €3 million budget and a remit to help secure raw material supplies for German industry. However, other countries in the EU, particularly those with a shrinking manufacturing base, had other priorities, and for the most part it is hard to avoid the conclusion that the EC initiatives in this area contained more form than substance. At most, the effect was to draw additional financial resources to existing programmes (such as R&D) which might be deemed to have a raw material dimension.

The matter of critical minerals nonetheless continued to be debated extensively within the European Commission and in the European Parliament and to be the subject of studies and expert group meetings. In May 2014, the EC published an updated version of its 2010 report, extending to 20 the number of minerals it now deemed critical.[12] It was agreed there was a need to monitor the position on raw materials and to engage in dialogue with raw materials producers, such as those in Africa, but there was little meaningful policy action beyond a challenge to China on its trade policies in the WTO (discussed below). There was a further 'communication' from the EC on the Raw Materials Initiative in 2011. However, it should be noted that a 'communication' is the weakest form of policy instrument available to the Commission. Following the global financial crisis (GFC) and the subsequent crisis around the Euro, there was little political will or money to do much with this agenda.

But, as tends to be the case with policy initiatives, once they are up and running, they have their own momentum.

The one area where the EU did enact new legislation pertaining to the mining sector had nothing much to do with domestic mining or with supply security. It related to the disclosure requirements of EU companies operating in the extractive and forestry sectors. With a view to encouraging transparency and thereby restricting the scope for corruption in the sector, in June 2013 the European Parliament approved a proposal requiring EU mining companies to disclose any payment over €100,000 made to local or central government in the countries where they were operating; such payments include royalties, licence fees and taxes. In so far as this requirement applied at project level, this initiative went well beyond the requirements of the EITI which only require disclosure at company level. According to the EU Commissioner responsible for the legislation, its intention was to 'bring in a new era of transparency to an industry which is far too often shrouded in secrecy...so both companies and governments can be held to account for the revenues from natural resources'.[13] Legislators brushed aside industry objections that the disclosure of this information would impose additional costs on companies and, in some countries such as China, Angola and Cameroon, put them in contravention of local laws.

US response

The United States was in very much the same position as the EU with respect to the issue of mineral raw material supplies. Estimates by the US Geological Survey and US Department of Commerce suggest that in 2013 mineral materials to the value of $689 billion were consumed in the United States in sectors of the economy accounting collectively for $2440 billion of value added. With local mine production declining or stagnating, an increasing proportion of these raw materials had to be imported. For many important commodities, ranging from fluorspar to manganese, the United States is 100 per cent import dependent. For others, including, potash, platinum, titanium minerals, cobalt and zinc, it has to import more than three-quarters of its requirements.[14]

Slightly ahead of the EU, a study was launched under the auspices of the National Research Council (NRC) of the National Academies to evaluate the criticality of minerals in the US economy. Using the same general framework as the EU study – with economic importance in use on one axis of the matrix and supply risk on the other – the researchers subjected 11 mineral commodities to detailed analysis. The mineral commodities were copper, gallium, indium, lithium, manganese, niobium, platinum group metals (PGMs), rare earths, tantalum, titanium and vanadium. On the basis of the analysis, the NRC determined that those falling into the 'most critical' zone of the matrix were indium, manganese, niobium, PGMs and rare earths.[15]

As noted previously, many minerals deemed critical are of particular significance to the energy sector. The rare earths neodymium, praseodymium and dysprosium are used in the production of permanent magnets employed in wind turbines and in motors for electric vehicles. Another rare earth, lanthanum, is used in advanced battery systems, as also is lithium. The rare earths europium, terbium and yttrium are used in energy-efficient fluorescent lighting systems, while indium, gallium and tellurium are used in thin-film photoelectric cells.

Accordingly, it was the US Department of Energy (DOE) which took the lead on the matter of critical minerals at an official level in the United States. In a report published in 2010, the DOE laid out plans for a programme of research, information-gathering and international dialogue with like-minded nations on the subject of critical materials.[16] It also advanced a strategy, like that in the EU, resting on three pillars: to encourage a diversity of supply for critical materials, including increased supply from within the United States; the development of substitutes for the materials posing the greatest risks; and increased recycling and reuse of critical materials. However, with respect to the first pillar, it noted that the matter of resource development in the United States was outside its departmental remit.

Following on from this report and from a second one on the subject which appeared in 2011, the DOE decided to establish an institute dedicated to progress its work on critical materials. In 2013, the Critical Materials Institute (CMI) was set up within the DOE's Ames Laboratory.[17] In line with the DOE's declared strategy, the mission of the CMI was to develop technologies, '(a) to diversify and expand the availability of these minerals throughout their supply chains (b) to reduce wastes by increasing the efficiency of manufacturing and recycling, and (c) to reduce demand by identifying substitutes for critical materials'. It was given a budget of $120 million over five years.

With respect to the matter of encouraging local resource development in response to concerns over supplies of critical materials, there was less progress and certainly no government-backed proposals to take the matter forward, this despite some outspoken comments from within industry that if the government was genuinely concerned about the security of mineral supplies, then this is something it should be addressing as a matter of urgency.[18]

As of late 2014, there were two bills addressing the matter of critical materials before Congress. The first, The National Strategic and Critical Minerals Production Act (HR 761), was aimed at boosting domestic mineral supply by streamlining the process for obtaining mining permits – its intent was to reduce the review process for mine permitting to 30 months – but it was opposed by the Obama administration because of concerns that it might weaken environmental protections. The second, the Critical Minerals Policy Act (S 1600), being considered by the Senate Committee on Energy and Natural Resources, envisaged a programme of up to $60 million for studies of domestic mineral resources and research into recycling and substitution. It proposed funding a review of the permitting process but did not mandate any changes. Testifying to the Senate Committee, the National Mining Association said, 'The United States has one of the world's greatest mineral repositories, but our ability to get these minerals into the supply chain . . . is threatened. [The main obstacle] is the length of time consumed in obtaining permits.' It went on to identify permitting delays 'as among the most significant risks and impediments to mining projects in the United States'. The United States is reckoned to have the world's lengthiest permitting process, averaging 7–10 years, compared with two years in Canada.[19]

As with the EU, the US government appeared generally to be rather less interested in what its mining companies could do to help alleviate supply shortages than it was in their standards of corporate behaviour. Under Section 1504 of the 2010 Dodd-Frank Wall Street Reform and Consumer Protection Act, it required mining companies with a US listing to disclose all payments to governments over $100,000. The provision, an exact parallel to that enacted by the EU at around the same time, was adopted by the US Securities and Exchange Commission (SEC) in 2012. However, following a challenge in the courts by the US oil industry and the US Department of Commerce, who argued that the provision would impose unreasonable costs on US companies and would in certain countries violate contract terms, implementation of the measure had to be deferred at least until 2015.[20]

A second provision, coming under Section 1502 of the Dodd-Frank Act, was designed (somewhat paradoxically in the context of the ongoing debate on critical minerals) to introduce new controls over minerals being imported into the United States. The provision required that companies buying certain mineral materials from conflict-ridden areas be able to demonstrate that these materials came from legitimate, licensed sources. This they would do by providing a certified paper trail leading right back to the mine. The provision, which followed in the footsteps of the Kimberley Process relating to conflict diamonds, was particularly targeted at tin, tantalum, tungsten and gold originating in the eastern provinces of DR Congo. This too was formally adopted by the SEC in 2012, although it too was subject to a challenge in the courts, this time by the National Association of Manufacturers and the US Chamber of Commerce. While the measure was formally implemented in June 2014, it was estimated that only 5 per cent of affected companies were compliant, most claiming that they had been unable to obtain the information in the form required under the law.[21]

After evaluating the operation of the US conflict minerals scheme, the EU decided not to follow the United States down this particular path but to opt instead for a looser voluntary scheme. In explaining its decision, the EU Trade Commissioner observed that the bureaucratic conditions of the US scheme and a fear of litigation appeared to have had the unintended consequences of driving US buyers away from the DRC to the cost of legitimate producers and their livelihoods and of encouraging smuggling.[22]

Rare earths case

On 7 September 2010, a Chinese fishing boat collided with two Japanese coastguard vessels near the group of uninhabited islands known as Senkaku in Japan and Diaoyu in China in the East China Sea. No one was hurt in the incident, but the Japanese coastguards arrested the captain of the fishing boat, Zhan Qixiong, and held him for two weeks. Given that ownership of the islands is disputed and the captain's conviction that he was operating in Chinese waters, this sparked a major diplomatic row. Towards the end of September, Japanese companies noticed that deliveries of rare earths from China had dried up. China denied it was withholding supplies of rare earths in retaliation for the incident, but it was nonetheless not until the end of November that supplies of rare earths to Japan resumed.

The obscure group of 17 mineral elements known as the rare earths (described in Chapter 3) became something of a lightning rod for concerns about critical minerals in the traditional mineral-consuming countries and for concerns about China's role in the supply of such minerals. It was also the one area where the traditional consumers had a clear and common strategy to address the problem.

Between the mid-1960s and the mid-1980s, the United States had dominated the world market for the rare earths, its production based on the Mountain Pass mine in California (Figure 8.2). The other major

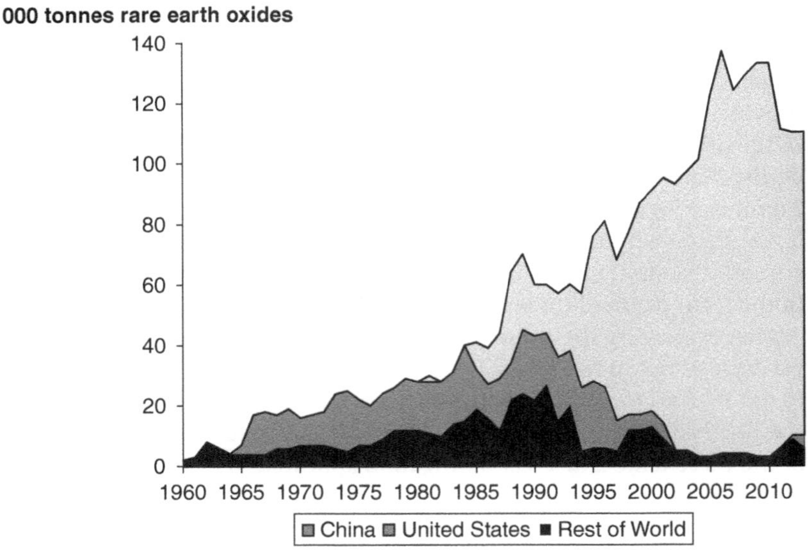

Figure 8.2 World production of rare earths, 1960–2013
Source: USGS and USGS (2011).

source of supply was Australia where rare earths were recovered as a by-product of mining mineral sands. However, in the mid-1980s, China started to export rare earths. As its production and exports grew, prices of rare earths declined. By 2002, as a result of the combination of low prices and environmental pressures (the company had been obliged to close its processing plant in 1998 as a result of wastewater leaks), Mountain Pass could no longer compete and was forced to close. Production in Australia also shut down. Since Chinese exports continued to grow and keep prices at subdued levels, consumers of rare earths were not too unhappy with this situation. In the mid-2000s, however, circumstances began to change.

With China's own demand for rare earths growing fast, it began in 2004 to discourage their export using a combination of export quotas and tariffs. As described in Chapter 3, the constraints on exports ramped up gradually until 2010 when China announced it was cutting export quotas by 40 per cent, just as global demand was recovering after the GFC. As the year progressed, prices of rare earths began to firm and then to fly. The export price of lanthanum oxide, which had started 2010 at around \$5,500 a tonne, had advanced to \$9,765 a tonne by July before soaring to a peak of \$153,560 a tonne in July 2011. The issue caught the public imagination and became the subject of numerous press stories. It also set alarm bells ringing in the capitals of the traditional mineral-consuming countries. The claim was that China's actions were part of an industrial policy designed to give Chinese manufacturers using rare earths an advantage over foreign competitors; moreover, this was in contravention of its obligations under the rules of the WTO and the terms of the accession treaty which China had signed when it joined the WTO in 2001.[23]

China's take on things was, understandably, different. The availability of adequate raw materials was viewed, as discussed in Chapter 3, as essential to its project for economic and industrial development and its movement up the value chain. However, as a rapidly developing country, China had many potentially conflicting policy objectives. As well as wanting to ensure adequate supplies of raw materials for its manufacturing industries, it was also seeking to raise the environmental and health standards of mineral producers and to better regulate the development and depletion of its remaining mineral reserves. There was a strong conviction that it was running through its domestic resources at too fast a pace. Export restrictions were seen as a pragmatic means of enabling it to meet these multiple objectives. Moreover, China claimed that the right to use its domestic resources to promote its industrial development and to determine how much of its mineral raw materials were made available to overseas buyers was a sovereign matter.[24] China was not an energy-rich country, nor a particularly mineral-rich one, and it did not see why it should shoulder the responsibility for supplying the world with energy-intensive and other mineral products, which could just as readily be supplied by other countries.

The General Agreement on Tariffs and Trade (GATT), which are the rules governing international trade, and of which the WTO is the guardian, had always been a bit vague on the matter of export restrictions. When the GATT was originally devised in the aftermath of World War II, the negotiators were principally focused on import restrictions, since the widespread use of such restrictions was the major barrier to free trade at the time. The GATT accordingly had relatively little to say on the matter of export restrictions. The major instance of export restrictions in the post–World War II world was the restriction imposed on the export of oil by the Organisation of the Petroleum Exporting Countries (OPEC), and oil fell outside the remit of the GATT.

Uncertain of how the GATT rules would be applied to restrictions on mineral exports, and indeed how these would mesh with the undertakings made on these matters by China at the time of its accession to the WTO, the United States and the EU, supported by Mexico, in late 2009 brought a case against China in the WTO with respect to its export restrictions on a group of nine mineral commodities, refractory bauxite, fluorspar, coke, magnesium, manganese metal, silicon metal, silicon carbide, yellow phosphorus and zinc. Although a very heterogeneous group of commodities, what they had in common was this. They were all commodities which China had in previous years exported in significant quantities, and they were all commodities where China's exports were now declining and where prices had been increasing.

The question to be addressed by the WTO panel hearing the dispute essentially boiled down to an assessment of China's motives in imposing export restrictions on these products. Although the default position of the GATT is that all export restrictions are prohibited, it does allow for the possibility of export restrictions in cases where countries can demonstrate that these restrictions have been 'temporarily applied to prevent or relieve shortages', are 'necessary to protect human, animal or plant life or health' or 'relate to the conservation of exhaustible natural resources'.[25]

These three provisions constituted the basis of China's defence in the case being brought. In its ruling, delivered in July 2011, the WTO Panel acknowledged that while these factors may have been part of the motivation for China's export restrictions, China had failed to demonstrate to its satisfaction that they were the primary reasons. Moreover, in view of the specific undertakings which China had made on export restrictions in its accession treaty when it joined the WTO (it had been permitted to retain export restrictions on 84 listed products on the understanding that it would not add any more), it was doubtful whether China was entitled to make use of the exemption provisions for export restrictions contained in the GATT anyway.

Encouraged by this outcome, the United States, the EU and, this time, Japan, brought a case against China on its rare earths export restrictions in March 2012. China once again advanced as the basis of the defence for its measures the need 'to protect human, animal or plant life or health' and to conserve its resources. Aside from providing evidence on the environmental destruction wrought by the mining of rare earths and the limited extent of its rare earth resources, China was able to provide substantial evidence of its attempts to bring better regulation to an industry it acknowledged had historically grown in a largely uncontrolled fashion. It also claimed that, since the ramping up of restrictions on exports had been accompanied by a similar ramping up of restrictions on the production of rare earths, export restrictions could not be said to discriminate again foreign buyers. It pointed out that, for all the complaints by foreign consumers about China denying them rare earth supplies, the export quotas on rare earths had never actually been fully taken up.

The panel duly ruled in March 2014 against China on the grounds that, under the terms of its accession treaty, China could not use in its defence the designated exemptions for the imposition of export restrictions provided for in the GATT. However, it also stated that even if China was eligible to benefit from these exemptions, it had failed to demonstrate that its quotas and duties were necessary to meet its environmental objectives and to conserve its resources, or that the measures it had in place were evenhanded in their treatment of domestic and foreign buyers of rare earths. Instead, the panel claimed, China's export quotas 'were designed to achieve industrial policy goals rather than conservation'.[26] Accordingly, China was out of compliance with the GATT and should remove its export restrictions on rare earths.

Although on one level clearly a victory for the United States, the EU and Japan, in practice, the outcome of the case seemed likely to change very little or to contribute materially to increased security of supply of rare earths. China announced that it would not allow the ruling to distract it from its efforts to regularise and clean up its rare earths industry and tighten controls over the rate of exploitation of its rare earth reserves. Moreover, by the time the panel reached its conclusions, the market conditions which had prompted the case had largely evaporated. After peaking in 2011, the prices of most rare earths elements had collapsed as consumers ran off the stocks they had accumulated during the earlier period of panic buying and as the effects of substitution began to take effect. The pressure on supply was also beginning to be relieved by the start-up of rare earth projects outside China. Mountain Pass in the United States was reopened by owners Molycorp in 2011 as was the Mt Weld rare earths mine owned by Lynas Corp in Western Australia. Other projects were being lined up in Canada, Australia, South Africa and Sweden. Most market analysts took the view that the WTO's ruling would have a negligible impact. Others argued that the results could well turn out to be perverse in so far as any increased supply from China would threaten the existence of the new non-Chinese production thereby reinforcing China's market dominance and reducing supply diversity.[27]

Anything but mining

The two actions at the WTO constituted the most decisive response by the United States and the EU to concerns over supplies of mineral raw materials. They may not have achieved much in practical terms, but they represented a clear assertion of the view that the best way to organise the global supply of minerals is through free markets operating within a rules-based framework. And they laid down a challenge to China to play by these rules.

Generally, however, the policy responses of the traditional mineral-consuming countries to the supply threats thrown up by the commodities boom lacked substance and coherence. Certainly, they paled into insignificance beside the robust and single-minded policies adopted by China to secure its mineral raw materials. Possibly, this was because, for all the public expressions of concern about the problem, deep down they were not convinced it was a serious one or else (as was effectively the case in the EU) they lacked the authority to tackle it. Given the prominent role the US Department of Defense (DoD) played in previous scares about the security of supply of minerals, it was noticeable how little it had to say publicly on the subject this time around.[28]

Another part of the explanation for the limited nature of the policy response was that mentioned earlier, timing. By the time the matter of mineral supply had forced its way onto the policy agenda, the boom was already turning to bust. Demand for mineral commodities collapsed in the traditional mineral-consuming countries in the second half of 2008, and governments faced a rapid re-ordering of priorities as they grappled with a sharp economic slowdown, massive debts and a shortage of cash. By the time of the second commodity price surge in 2011, the investments stimulated by the first price boom were beginning to show, through in the form of new mine capacity, and the whole issue was beginning to lose its urgency.

A third possible reason for the lack of policy coherence was that it was the result of a lack of understanding in government circles about how mineral markets work and where exactly the threat to those markets lay.

Some of these problems stemmed from the simplistic analytic framework employed to evaluate the criticality of minerals. As noted, this framework was pretty much unchanged from the framework developed to address similar concerns during the late 1970s and early 1980s when South Africa and the Soviet Union rather than China were the primary focus of concern about mineral supplies. The notion of 'economic importance in use', which this framework used as one of its criteria for assessment, is economically flawed. All mineral substances are 'important' in one respect or another or they wouldn't be employed, but the assessment of *relative* importance is deeply problematic. Many of the minor metals assessed as critical are deemed to have high importance in use, but common sense suggests that more mundane high-volume commodities such as copper and iron are ultimately more important to the fabric of modern industrial civilisation. The ranking of rhenium above copper and graphite above iron in the EC's 2010 assessment appears anomalous and a questionable basis for devising good policy.[29]

The second axis of the critical materials framework – that relating to supply risk – was not without problems either. As a research paper produced under a programme funded by the EC pointed out, supply risks come in many different forms and accordingly require different solutions. Specifically, the authors of the paper pointed out that the framework used by the EC failed adequately to distinguish between short- and long-term supply problems, took insufficient account of the specific characteristics of the different resource markets analysed and disregarded the role played by the production chain – that is to say, the refining and transport of minerals and trade in semi-fabricated products – in its assessment of mineral supply.[30] With respect to this last point, some of the minor metals entering the EU come in as minor constituents of the ores of other metals and thus effectively depend on Europe continuing to be a major smelter and refiner of metals. Most of Europe's cobalt, for example, enters as a minor constituent of copper and nickel ores. The importance of supply chains is also highly pertinent to the evaluation of the situation in rare earths in so far as China's dominance in this market arises every bit as much from its control of rare earth processing technologies as it does from its possession of rare earth resources in the ground. In short, these are matters which cannot be reduced to a single vector.

The analytic approach adopted in the evaluation of critical minerals led naturally, and possibly deliberately, to a policy focus on a handful of exotic but economically relatively minor minerals originating in one country. One reputable consultancy flagged up the danger that the approach risked leading to 'scaremongering' among audiences which did not understand the limitations of the methodology and might potentially prompt unnecessary substitution.[31] Others questioned whether it was either economically efficient or the best use of scarce innovative talent to have governments subsidise research into trying to find ways of not using materials which are known to do a job well and can in no wise be considered scare in nature.[32]

A further consequence of the particular focus given to the debate on critical minerals was that it drew attention away from the broader systemic challenge to the supply of all mineral raw materials posed by the rise of resource nationalism and growing state intervention in mineral markets, as discussed in the Chapter 7.[33] Defining 'critical minerals' in such restrictive terms had the inevitable consequence of implying that other minerals were not critical. It also had the effect of making China the principal target of the policy response, thereby overlooking the considerable areas of overlapping interest which traditional mineral-consuming countries have with China as a large consumer and importer of minerals. The policies adopted by Indonesia with respect to its export of minerals, for example, are of relevance to all mineral-consuming and mineral-importing countries. Finally, it resulted in a policy focus, which gave primacy to trade measures and to programmes of demand management (e.g., policies focused on the efficiency of use of minerals) rather than to the cultivation of new supply.

Logic suggests that an obvious way to address concerns about mineral supplies is to encourage mining investment in safe and secure places. Although policy initiatives on both sides of the Atlantic paid lip service to the importance of domestic mineral production, there was little practical attempt to try to reverse the declining trend in production of recent years. On the publication of the EC's study on critical minerals in 2010, the editor of one trade journal remarked, '...perhaps the EC could take a look in its own backyard. One thing this exercise might accomplish is a re-evaluation of the EC's stringent legislation for companies trying to open up new sources of minerals *within* the EU, and thus for the benefit of EU consumers – a constant source of frustration for many EU mineral developers.'[34]

Nor was there any greater enthusiasm to put in place measures to encourage mining investment overseas. No interest was shown in supporting or encouraging mining companies in their activities outside their home regions, in fronting government-to-government deals under which companies might operate, bringing pressure to bear on the World Bank to promote mining-linked infrastructure deals of the sort the Chinese were engaging in, or lending diplomatic support for listed companies where the legal rights of these companies were being infringed. Nor were there any evident attempts to engage with mineral-rich countries having a predisposition to resource nationalism in a debate on how the interests of both producing and consuming countries could be satisfied within a changing geopolitical context. The only point at which international mining companies appeared on the policy agenda was in the largely negative context of their governance.

It is possible that this stance of the traditional mineral-consuming countries reflected in part the almost-complete absence of global mining companies now based out of continental Europe and the loss of

US leadership of the global mining industry. Relying on the operation of mineral markets for sourcing raw materials is, of course, a perfectly reasonable stance to adopt so long as everyone is playing the same game. But the signs are that this is less and less the case. The centrepiece of the policy response of the traditional mineral-consuming countries appeared to be the hope that the rest of the world might be cajoled back to a free market, rules-based system for the supply of mineral raw materials. But the geopolitical system which sustained and policed this system – essentially the unipolar US-dominated world of the Washington consensus – was gradually eroding. This being the case, the policy responses adopted looked inadequate to the task in hand. It is a challenge for countries to insist on others supplying them with the raw materials they are transparently disinterested or unwilling to supply for themselves.

9
The Next Thirty Years

Where is mining headed?

In the Pilbara in Western Australia, automated driverless trucks the size of houses relentlessly plough backwards and forwards in the vast open pits of iron ore overseen by operators 1500 kilometres away in Perth. Almost half of the uranium mined in the world today is recovered using solution mining. This involves leaving the ore in the ground and recovering the minerals from it by dissolving them and pumping the resulting solution to the surface where the minerals can be recovered. In Yorkshire, United Kingdom, there are plans to build an underground potash mine in a national park. The winding gear for the mine as well as the maintenance shops and ventilation equipment will be located underground so as not to be visible at the surface. Canadian company Nautilus has an advanced stage project to recover copper, gold and silver from sulphides at a depth of 1600 metres on the floor of the Bismarck Sea off the coast of Papua New Guinea and is investigating the possibility of recovering polymetallic nodules at around 4500 metres depth in the Central Pacific west of Mexico. In 2012, a company named Planetary Resources announced proposals for mining near-earth asteroids (NEAs) for precious metals.[1]

It is impossible to be sure what the mining industry will look like in 30 years time, although asteroid mining looks a long shot. What one can be sure of is that it will be different from today, and different from what most people today expect.

In the conclusions to his book, *Trouble in the Third Kingdom*, Simon Strauss observed that the situation faced by the mining industry in the mid-1980s bore little resemblance to the conditions being predicted for the mining industry 30 years before. Concerns over mineral shortages had been displaced by the crushing reality of abundance. In looking forward over the next 30 years, Strauss expected demand growth to be subdued. He took a dim view of the prospects of developing economies while China didn't even rate a mention. He supposed that 'advances in the technology of exploration can be expected to lead to the discovery of abundant supplies of most major minerals for many years to come'. He thought increased trade between East and West would be good for growth but saw little likelihood of it occurring. Mining would, he asserted, 'remain a risky proposition' and mineral enterprises would only be able to secure access to sources of capital 'if they conduct themselves in a manner that engenders satisfactory credit ratings'.[2]

Things did not work out the way Strauss predicted. Such, unfortunately, is the fate of most forecasts. The collapse of the Soviet Union and the revolution inspired in China by Deng Xiaoping brought the 'East' into the global mainstream in a way that could not easily have been foreseen. The opening up of world markets for goods and capital galvanised growth in the emerging economies, enabling them to play a much bigger part in global growth and in the demand for minerals than seemed likely in the mid-1980s. The outlook for mining may have looked bleak from Strauss's perspective in the United States, but mining companies elsewhere in the world, both in the West but increasingly also in the East, were stepping into the breach to assume a leadership role. Investors, who had earlier been persuaded by the notion that mining was a deadbeat industry condemned by Western de-industrialisation, woke up to the reality that mineral raw materials were an essential ingredient for the development of emerging economies as well as a critical input to the most sophisticated technology products. Iron ore and rare earths had suddenly come to look a lot more interesting.

This final chapter briefly considers the state of the commodity cycle before turning to the industry's longer-term prospects. It considers the factors likely to shape global mineral demand over the next 30 years and some of the key drivers of prices and production costs, including the effects of mineral depletion. It then looks at some of the 'above ground' influences on mining investment, in particular environmental pressures on the industry and political constraints on investment. It concludes with some thoughts on what these trends might imply about the future evolution of the mining industry.

The cycle turns

The near-term outlook for the mining industry is, as ever, dominated by cyclical influences. The tidal wave of investment, which poured into mining during the boom years, put output for many commodities on a strong upward trajectory. At the same time, conditions of demand have been suppressed by the ongoing travails of the debt-ridden advanced economy countries and the slowing of China's economy. Global growth, which rebounded over 5 per cent in 2010 following the global financial crisis (GFC), has since settled back to around 3 per cent a year. Markets have moved into better balance, and some of the exuberance has left the commodity space. By late 2014, the prices of many mineral commodities had fallen back to where they were five years previously, in the aftermath of the GFC. Investors had turned cool on the mining sector, and exploration companies were struggling to attract funding for their projects.

As commodity prices have softened, mining companies have been obliged to adjust to the new realities. Industry leaders have rediscovered old truths about the economics of the industry, the importance of cost control and of good capital management in a capital-intensive industry, the importance of free cash flow and the importance of keeping a focus on the long term in what is by nature a long-term business. They have also rediscovered the importance of rewarding investors and have sought to regain their trust by boosting dividends and implementing share buybacks. One might reasonably ask how these lessons came to be forgotten over such short period of time. But they did. Ultimately, success in mining is about the efficient deployment of capital and not the anticipation of mineral demand or the building of corporate empires. Companies need to control what they can control – their capital spend and their costs – and take the market as it comes.

Given its over-riding significance to global demand, where things go next depends heavily on what occurs in China. In retrospect, it is clear that the growth surge in China, which ignited commodity markets over 2005–2011, was a unique historical event. It reflected China's establishment of itself as the workshop of the world. It was not the beginning of a sustainable long-term trend as some believed. China will continue to grow but at a much reduced pace. Over-investment in industry and in housing has left the country with too much built capacity, and demographics are working against future growth. From 2015, the working population of China is set to decline. Moreover, and as important as the slowdown in growth itself from the perspective of the minerals industry, a rebalancing of the economy away from exports and investment towards domestic demand and consumption will result in a less material-intensive form of growth.

Reduction of the share of investment in gross domestic product (GDP) in a controlled and gradual fashion is not something easily achieved.

Most countries which have been required to make this adjustment have done so by way of an economic crisis: Japan in the mid-1970s, Korea, Thailand and Malaysia in the late 1990s. China, having greater control over its banks and external capital flows, enjoys certain advantages in managing the process that these other countries lacked, but, even so, it is hard to see how the transition can be achieved without significant loss of economic momentum. This is already occurring. Growth in China's industrial production in 2014, at 8 per cent, was at its lowest rate since 1991. President Xi's anti-corruption drive was reported to be leading to paralysis of decision-making in some key economic sectors such as power generation. Property sales and property prices have declined and the rate of investment in real estate was expected to turn negative in 2015. Not unrelated to this, apparent consumption of crude steel in China, according to the China Iron and Steel Association (CISA), declined in 2014 for the first time since 1995.[3] Reflecting the combination of weakening domestic demand and still-growing metal production, China's exports of steel and aluminium products were trending upwards.

What happens to mineral demand in China is, in part at least, no more than the normal ebb and flow of the commodity cycle. Fluctuations in demand are invariably the prime driver of price cycles in mineral markets,[4] and China is a big element in that demand. The purpose of this chapter, however, is not to speculate on the course and timing of the cycle but rather to consider the context in which the cycle is played out, focusing on the things in the mining industry which have changed and are likely to change, with respect to what is produced, where it is produced, how it is produced and who produces it. As the industry relapses back into a slower-moving and lower-profile mode of operation, it may appear simply as if things have come full circle from where they were in the early 2000s. And in one sense this is true. But it is the premise of this book that beneath the natural fluctuations of the commodity cycle, the mineral industry has changed and changed in ways that are unlikely to reverse.

Long-term demand trends

One of the key changes that has taken place in recent years is the customer base of the mining industry. Twenty years ago, the primary focus for mining industry analysts trying to divine the likely direction of their markets was economic trends in the advanced economies. Even in 2000, between one half and two-thirds of major metals use was taking place in advanced economy countries (Figure 9.1). In 2013, the shares accounted

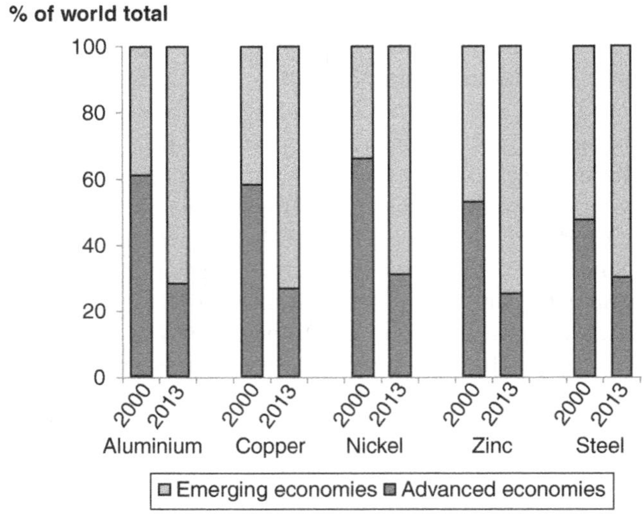

Figure 9.1 Distribution of global mineral consumption, 2000 and 2013
Source: WBMS, WSA.

for by this group of countries had halved to between a quarter and a third. For all of the major commodities, consumption in the advanced economies in absolute terms declined through this period. Partly, of course, this reflected the relocation of basic industrial production from the advanced to the emerging economies, but it also reflected the faster rates of economic growth in the emerging economies and the faster growth in their own domestic demand.

It is likely that this will be the pattern of the future. There remain huge development needs and huge development potential in emerging economies. With their higher underlying rates of economic growth, it must be supposed that they will account for an increasing share of global output and an increasing share of the global demand for minerals. Economic growth and minerals demand in the advanced economies will continue to be inhibited by their ageing populations, problems of indebtedness and their diminished competitiveness in basic manufacturing.

A key driver for commodity use within the emerging economies will be the rapid growth in the urban middle class. As discussed in Chapter 2, industrialisation and urbanisation advance hand in hand. And, just as industrialisation implies a higher use of mineral commodities, so too do urbanisation and rising living standards. Growing urban populations require better, more material-intensive housing, more domestic appliances and better supporting infrastructure – power supplies, water services and transportation.

In a study published in 2012, McKinsey Global Institute forecast that the largest 440 cities in the emerging world would deliver half of global GDP growth between 2010 and 2025.[5] Over this same period, it forecast that the consuming classes, defined as those with incomes of over $3600 a year (in 2005 PPP dollars), would expand from 2.4 billion to 4.2 billion people, raising the share of this group in the global population from 36 per cent to 53 per cent. Cities in the emerging economies would account for 95 per cent of the increase. It projected an even more rapid growth in the numbers of the higher-earning middle classes. In the same 440 cities, it expected households with incomes of over $20,000 a year to increase from 114 million to 296 million, an annualised increase of 6.5 per cent a year. All of this points to the requirement for a lot more minerals.

The combination of the faster growth of the emerging economies and the more material-intensive nature of this growth was a key factor underpinning the commodity boom of the 2000s and a key part of the thinking which led super cycle theorists to maintain the world was looking at an increase in the trend growth rate of mineral demand.

While the growth of the urban middle classes will undoubtedly be a positive driver for future mineral demand, this does not mean it will be transformational in quite the way in the super cyclists believed. As was shown in Chapter 2, evidence of acceleration in the rate of demand for mineral commodities over the past decade or so is mixed. While there was clearly acceleration in the demand for constructional products and their raw materials (steel, iron ore, coking coal, cement and aluminium) during China's period of intense industrialisation, other commodities did not experience the same acceleration in demand at a global level. Moreover, it cannot be assumed that the faster growth rates in the demand for these constructional raw materials will be sustained as China's growth model changes and the constraints on growth imposed by environmental conditions in China, notably water shortages and poor air quality, intensify. It is also likely that, as China's economy matures, more of it raw material requirements will be met from domestically generated scrap. In sum, future growth in China will be less material intensive.

As the rate of China's growth slows, the baton of mineral demand growth will pass to other parts of the emerging world. Collectively, there is a vast reservoir of unfulfilled demand in these countries.

India currently under-consumes mineral raw materials just as China's over-consumes them. Although recent growth in India has been hampered by overweening bureaucracy, poor infrastructure and rampant corruption, the country's young and growing population and its strong enterprise culture mean it unquestionably has the potential to sustain rapid economic growth during the next 30 years, given the right policies. And there is another wave of populous emerging countries with the potential to grow explosively coming along in support. This wave includes a grouping referred to by the Goldman Sachs team who invented the concept of the BRICs as the 'MINTs', signifying Mexico, Indonesia, Nigeria and Turkey.

These other emerging economies are, however, unlikely to follow China's growth trajectory. China's centrally directed model of development, with infrastructure being built out ahead of the need for it, is transparently not being adopted by emerging economies such as India. Moreover, in smaller countries without the same emphasis on state support and without access to cheap finance such as that which Chinese enterprises have enjoyed, there is likely to be a more cautious approach to the development of metal projects and raw materials use. For example, private companies looking to establish or increase steel production in some emerging economy countries may be reluctant to go directly to the construction of costly integrated steel mills using large amounts of iron ore and coke and choose instead to build smaller electric arc furnaces employing steel scrap and direct reduced iron as their feed. In short, it is likely that the Chinese experience – with its huge economy and unprecedented 50 per cent of GDP going to investment – is an unreliable guide to what the future holds for other countries.

Equally, it cannot be supposed that countries will follow the consumption patterns of the advanced economies. Countries coming late to industrialisation do not have to pass through all the phases of development that those going before them did. They can leapfrog to more effective, resource-efficient technologies. Some emerging economies have gone straight to mobile telephone networks and will not need to establish the landline networks still widely employed in advanced economies. Moreover, they can learn from the experiences of those who have gone before. It is not uncommon in looking at the car market to measure a country's development against car usage in the United States. But the car market in the United States reflects some special historical and geographic factors, not least the long-standing availability of cheap oil and lots of cheap land. Many emerging countries will not be in this position. Moreover, they will have the experience of other countries, which have strangled their cities with urban traffic and which are belatedly trying to improve their mass transit systems, to learn from. In fact, as a general observation, economic growth in future everywhere is likely to have to face greater constraints arising from environmental considerations and a far greater emphasis on resource efficiency than in the past.

Accordingly, while there are good grounds for believing that demand for most minerals will continue to underpin growth in the mining industry, those hoping for a sustained uplift in the trajectory of demand growth are likely to be disappointed. After their rapid economic convergence with the advanced economies in the first decade of the century, the rate of growth in emerging economies has slowed. A slowdown in the growth of global trade and in cross-border investment is not helping. Moreover, growth in emerging economies will probably have to offset the drag from what could continue to be flatlining or declining demand for minerals in the advanced economies. As the euphoria of the commodities boom dissipates and the industry adopts a spirit of business-as-usual, analyst expectations have moved back closer into line with the historical experience. Naturally, there will be some variation by commodity. In view of the infrastructure and housing requirements of emerging economies, growth rates in the demand for construction raw materials may not fall back to pre-boom levels. Moreover, several technology metals should enjoy strong secular growth as the technology sector grows ahead of industrial output more generally. Rare earth usage, for example, is projected to grow at around 6 per cent a year.[6]

What the continuing growth in demand for mineral raw materials does mean, however, is that the *absolute* quantities of minerals required will continue to grow. This is highly relevant for an industry where the limited nature of the resources being exploited precludes a supply response based on the exploitation of ever-increasing economies of scale. Very large ore bodies are scarce, and few new ones have been turned up in recent years despite high levels of exploration. Accordingly, while there will doubtless be scope for expansions at mines where the scale of the ore body will support it, growing demand for minerals means the need for more mines. Consider the case of copper. A copper mine with a production of 200,000–250,000 tonnes per year of contained copper can be considered a large mine. There are relatively few copper deposits in the world large or rich enough to support this level of production. Among those falling into this category are, for example, Bingham Canyon and Morenci in the United States, Andina and Collahuasi in Chile and Cerro Verde in Peru. Assuming for the sake of argument a growth in demand for copper mine product of 2.5 per cent a year and a rate of depletion of another 2.5 per cent a year (reflecting declining grades and mine closures due to exhaustion of economic ore), as of 1990, the world would have needed two new large mines a year to meet the expansion of demand. By 2012, this requirement had doubled to four. And, as Mark Twain remarked when recommending the purchase of land, nature is not making any more.

The other thing that the changing pattern of global demand is bringing with it is a shift in the centre of gravity of metal trading. It is also bringing changes in the way in which mineral products are traded and the manner in which they are priced.

With regard to metal trading, this is experiencing a decisive shift towards Asia. For a long time, the metal markets have been focused on London, in large part because of the presence there of the London Metal Exchange (LME). LME warehouses, which are normally located in areas of metals consumption, have traditionally been in Europe, the United States and Japan. Gradually, the LME has extended its reach in Asia with warehouses being established in Singapore, South Korea, Taiwan and Malaysia. In 2012, the LME was itself acquired by the Hong Kong Stock Exchanges and Clearing (HKEx). Under its new ownership, the LME is seeking to extend further its warehousing system in Asia, develop the use of Asian benchmark prices and introduce contracts specifically tailored to the Asian market, the latter to include new, smaller contracts ('minis') aimed at Chinese mainland customers and cleared in Renminbi (RMB). Since 2011, the LME has supplemented its traditional October LME Week in London with an LME Week Asia, an event which is drawing business and attention away from London.[7] This comes, as noted earlier, as large Western banks are withdrawing from the physical metals trade.

With respect to the manner in which business is conducted and how commodities are priced, it tended to be the case that when the advanced economies dominated the trade in mineral raw materials, the business had a strong contract-based, long-term emphasis. Stability was seen as key to business success and institutional relationships were cultivated over long periods. The iron ore and coal businesses in Australia were founded on this model, and on co-operation with large corporate customers in Japan, Korea and Europe. In the world of the future, in which China and India will play a much more prominent role, while some vestiges of the old system will remain, much more business will be transacted opportunistically on the basis of spot cargoes and spot pricing. The nature of the customer base will be less institutional and more commercial, as discussed in Chapter 4. Very likely it will see greater emphasis on smaller family-owned firms and on relational business. Accordingly, mining companies making investments will have to make a judgement call on the market conditions they are likely to face without the security of long-run contractual commitments that they enjoyed in the past. This will create challenges in the evaluation of projects but may also have implications for the ability of these companies to raise finance.

Whither prices and costs?

Important as changes in the customer base are likely to be to future mining investment, the challenges faced by the mining industry in managing price and market risk go a whole lot wider and deeper than this.

Chapter 4 presented reasons why it is unlikely that real US dollar mineral prices will return to where they were at the start of the cycle and why the long-run downtrend in real mineral prices is not going to be restored any time soon. These reasons related to a shift in the terms of trade between the prices of commodities and other tradable goods and a secular decline in the value of the US dollar. The rupturing of the downward trend in real prices will, however, provide no comfort to mining companies since these companies are interested in net prices – prices less costs – and costs will have been similarly affected by these two factors. Margins are returning to pre-boom levels even if prices are not.

The bigger question with which companies must grapple is how much of the upward pressure on costs experienced during the boom years will prove cyclical, and therefore reversible, and how much will be sustained into the future. Much hinges on this judgement. Mining companies commonly look to industry cost curves to derive the long-run prices they use in their investment decisions. Their assessment of these long-run prices rose during the boom years as cost curves lifted and steepened. And these curves lifted and steepened a lot. Between 2004 and 2011, it was estimated by commodity consultants CRU that the marginal cost of producing copper rose at an average of 13 per cent a year in real terms, effectively more than doubling over the period.[8] Between 2003 and 2013, it has been estimated that nominal cash costs of the world's ten largest gold producers increased 14 per cent a year, a more than tripling.[9] Although the prospect is for cost curves to lower and to flatten, there is much uncertainty about where they will settle out. Accordingly, the range of uncertainty regarding future prices is also considerable. During the 1990s, while views on long-run prices of commodities naturally differed, they nevertheless displayed a relatively high degree of convergence. This is far from being the case now.

The question of how much downward flexibility cost curves have when markets move into oversupply and where therefore prices will find a floor can only be answered with certainty after the event. Unfortunately, companies have to make decisions in advance of possessing this knowledge. If the history of the 1980s and 1990s is any guide, then conditions of oversupply will cause cost curves to fall further than many in the industry would like to suppose, and probably stay down for longer too. Producers only really get creative about cost reductions when they have to, and cost curves have in the past proven a rather soft 'floor' for prices in weak market conditions. As prices fall, so management's leverage relative to their suppliers, their workforces and governments tends to increase, giving them scope to push costs lower. At the same time, a more concerted push on work practices and the application of technology can bring about cost-reducing improvements in productivity.

Iron ore, a commodity which was a particular beneficiary of the boom and which contributed substantially to increased profitability of the global diversified miners (GDMs), illustrates the point. As described in Chapter 3, iron ore experienced a steep rise in its cost curve as higher-cost capacity in China, India and a whole raft of other smaller producers was drawn into production to meet the market's growing requirements. Gradually, however, as high prices persisted, new, large-scale, low-cost production was lined up in Australia and Brazil. In Australia, the largest producers, Rio Tinto, BHP Billiton and Fortescue added substantial new capacity and a sizeable new player, Hancock Prospecting, announced plans to enter the business. In Brazil, a large amount of new capacity was committed by Vale and Anglo American. Over the course of 2014–2016, BHP Billiton has estimated that something of the order of 400 million tonnes a year of seaborne iron ore capacity will start up, an amount equivalent to around 25 per cent of 2013 supply.[10]

That this new low-cost capacity will drive some higher-cost producers out of business is not much in doubt, but there is plenty of scope for debating how long this will take and how severe will be the impact on prices. It is likely that some 200 million tonnes a year of China's higher-cost production will be displaced along with the exports of many smaller producers elsewhere in the world, such as Mexico, Malaysia and Iran, who helped plug the gap in the market for seaborne iron ore when large producers in Australia and Brazil were unable to supply. Many of the iron ore projects being lined up in West Africa will be deferred as will smaller projects in Australia. High-cost producers hoping that the big producers would throttle back on their output in order to support prices face disappointment. There is little evidence in history of the industry behaving in this fashion. Big producers will seek to regain market share lost during the boom years when they were unable to expand their production quickly enough and will seek to optimise their margins by exploiting their economies of scale rather than by supporting prices. Determining who will be the marginal producer in five years' time and where therefore prices will tend is the key to future pricing. What is clear is that it will be a long way below the level of prices experienced during 2010–2013. At best, they will reflect the production costs at Australia's or Brazil's higher-cost mines. As one iron ore mining executive observed, 'Iron ore mining isn't tennis, it's a contact sport. Sometimes it hurts a bit . . . '.[11]

The tendency of individual mineral producers to pump up production during periods of oversupply may seem contrary to economic logic and a tactic liable to aggravate cyclical fluctuations in prices. But it is a widely observed feature of the mining industry, arising partly from the capital intensity of the industry and partly from the forces of competition. What appears to make sense from the perspective of the market as a whole does not always make sense from the perspective of the individual producer. In the face of falling prices, producers will commonly seek to cut their unit operating costs by squeezing out as much production as they can. In a capital-intensive business, this serves to spread fixed costs over a larger tonnage of product. In the copper industry, there is a clearly observable tendency for mine producers to overshoot their production targets in weak markets, just as there is a tendency to undershoot them in strong ones. Thus, between 1993 and 2001, a period of generally weak copper prices, global mine production exceeded forecasts in six of the nine years. Gold shows a similar pattern, although for slightly different reasons. When gold prices are high, output will commonly fall as gold miners seek to stretch out the lives of their reserves by exploiting low grade ores which might not be economically viable under weaker market conditions. The downward trending of world gold production in the decade following 2000 was a manifestation of this. By the same token, they may 'high grade' during times of price weakness. In other words, the tendency of gold producers to work lower grade ores is a product of higher prices as well as a cause.

While it may be safe to assume that most of the cost pressures experienced during the boom years will be cyclical and thus reversible, it is equally clear that there are some which do not fall into that category.

For example, a strong case can be made that the energy costs of mining will not go back to where they were before the boom. For an energy-intensive business like mining and mineral processing, this is a material consideration. There is also a strong case to be made that efficiencies arising from the application of increased economies of scale, which made a big contribution to cost reductions in the 1980s and 1990s, as noted in Chapter 1, will make less of a contribution in future. Mines are not getting bigger.

The question of what role depletion plays in supporting higher costs, or driving them higher, is harder to assess. The tendency towards lower grades in copper and gold, often cited as evidence of encroaching depletion, is more complicated than at first it may appear, as indicated above. Moreover, the fact that ore grades have a tendency to decline has not in the past represented a barrier to lower production costs. For most of history, advances in technology have been able to offset the effects of grade decline.

However, the notion that there are real physical pressures on the availability of mineral resources, which may constrain future production, goes a bit wider than grades. A lot of attention has been directed in the oil industry to the notion of 'peak oil', the idea based on the work of M. King Hubbert in the 1950s, that limitations on oil reserves mean the world is moving inexorably towards a point where oil production peaks, before going into long-term decline. As a result of the 2000s commodities boom, some analysts began to ask whether the same principle might not apply to other minerals, and to develop notions of 'peak copper' and 'peak lead', for example.[12] There are many reasons to doubt that this is so – fundamental differences between the geology of oil and other minerals and a failure to take proper account of the impacts of technology to name but two[13] – but the likelihood that the world is not in imminent danger of running up against physical resource limits does not rule out the possibility that the extraction of minerals will be more costly in the future than it has been in the past.

The extent to which depletion of high-quality, lower-cost resources might impact mining industry costs in future will vary according to commodity. There are vast quantities of good-quality iron ore around the world, even if their development may in some cases pose substantial infrastructure and political challenges. In the case of copper, however, exploration in recent years has failed to turn up large, rich, new deposits of the quality, for example, of Escondida in Chile or Grasberg in Indonesia. There is significant evidence that the scale and grade of copper discoveries is declining and that the cost of identifying new resources is increasing.[14] The average head grades for the world's copper mines fell from around 1.5 per cent contained copper in the 1980s to 1.1 per cent in 2010, while the amount of barren rock needing to be stripped in order to get at the ore in open pit mines gradually increased. Furthermore, as the better open-pit resources are worked out, more and more copper production in future will need to take place in underground mines, an inherently more difficult and costly form of mining. Estimates by Rio Tinto suggest that the proportion of global copper production supplied from underground mines will rise from 27 per cent in 2013 to 51 per cent in 2019.[15]

For nickel, it is apparent that many of the world's massive sulphide deposits, which have historically been the basis for the world's production of refined nickel, are depleting. Alternative supplies from lower-grade sulphide ores or from oxides (using high-pressure acid leach technology) are inherently more costly. For platinum and gold, the world's best ore bodies (or the best ones that are known about) are depleting and future production seems destined to come from smaller, lower-grade ore bodies, some of which pose greater technical mining challenges. The average grade of gold processed in the world's mines halved in the decade to 2013 to around 1.3 gram per tonne,[16] while it has been estimated that the cost of discovering a new ounce of gold increased five times between 2006 and 2013 from $16/oz to $87/oz.[17] In short, while the evidence may be inconclusive, there are grounds for supposing that, for some minerals at least, depletion will be a growing source of upward cost pressure for miners in the future.

Above ground issues – environmental

Only part of the resource development challenges which companies face flow from what is in the ground. The other part flows from what goes on above the ground, in the form of regulatory, environmental and political constraints on mining. In fact, many in the industry take a fairly benign view of the challenges of dealing with what is in the ground. These are generally practical problems, susceptible to technical solution. Above ground problems are often less tractable and more unpredictable.

A growing issue for the mining industry, as mine production grows and living standards rise, is the management of the environmental side effects of mining, which is to say mining's impacts on land, water and air.[18]

Historically, land use has been the most obvious way in which mining has impacted the environment. Mining has intense local impacts on the landscape, a reason why it attracts public attention and why it is heavily regulated. Mining often generates large quantities of solid waste as rock is removed to get at mineral-bearing ores and as the mineral in the ore is concentrated into a marketable product. For minerals like coal or high-grade iron ore (i.e., ore having an iron content of 60 per cent or more), where most of the recovered mineral is usable, the amount of waste can be relatively modest. However, for products where the mined grade of ore is less than 1 per cent, as is the case at many copper and nickel mines, hundreds of tonnes of rock have to be dug up to recover one tonne of metal, most of this comprising waste. In the case of gold, where grades are typically measured in grams per tonne (effectively, parts per million), the rate of waste generation is higher again. Across the mining and quarrying industry as a whole, it is estimated that around 52 billion tonnes of material are removed from the earth each year, with 40 per cent of this being waste. Of the 30 billion tonnes of ore extracted, about half of this comprises aggregates (stone) with the balance split fairly evenly between coal and metals.[19] Big as these numbers sound they are set to get bigger. Growing pressure on land use will, in all probability, add to pressures on the disposal of mine waste and, quite likely, its cost.

Water is used at every stage of mining and metal production. Its use is particularly heavy at the stage of mineral processing where ores are transformed into concentrates. Broadly speaking, water use in the recovery of metals correlates negatively with the grade of ore mined; the lower the grade, the greater is the water use. Coal washing is also very heavy on water. Although water use naturally varies enormously from project to project, it has been estimated that the production of a tonne of copper typically requires 172 tonnes of water, a tonne of nickel 107 tonnes and a kilogram of gold 716 tonnes. Over the years, the mining industry has become much more efficient in its use, and reuse, of water, and for certain processes has been able to employ seawater. It is nonetheless the case that some of the most mineral-rich areas suffer water scarcity. Northern Chile, a major copper producing area, is one of the driest places on earth. Western Australia, the largest iron ore producing area in the world, is water short. So is northern China, where much of the country's coal is produced. Addressing this challenge can be costly. BHP Billiton and Rio Tinto are investing $3 billion in a desalination plant for their Escondida mine in Chile. Significantly, in Ernst & Young's annual survey of risks facing the mining industry, water availability appeared among the top ten risks for the first time in 2014.[20]

Then there is the matter of airborne emissions. Metal production, as noted before, is an energy-intensive affair and, with energy use, often come carbon emissions. In some instances, carbon emissions also arise in metal production where carbon is used as a reductant, that is, a means of removing the unwanted oxygen from the mineral matrix. Using very broad averages, the production of a tonne of steel results in the generation of around two tonnes of carbon dioxide, a tonne of copper between four and six tonnes of carbon dioxide, subject to the technology employed, and a tonne of aluminium 20 tonnes of carbon dioxide. (In each case, it is assumed that electrical power is generated by burning hard coal. Where hydroelectricity is used to produce aluminium, emissions are nearer ten tonnes of carbon dioxide.) Clearly, any serious attempt to impose a charge on carbon emissions as part of a move towards a lower-carbon world must impact negatively on metal production costs.

As population grows, the call on the minerals industry will increase. And as the industry expands to meet demand and is required to work lower grades of ores and remove more overlying materials to get at the ore, the waste and emissions from mining will increase. At the same, as living standards rise, it is likely that the sensitivity of the public to the condition of their natural environment will grow. As discussed in Chapter 8, in some of the richer parts of the world, there is already clear evidence of revulsion against mining and a lack of preparedness to tolerate the activity at almost any price. This is commonly accompanied by a conviction that increased efficiency of resource use could significantly reduce the need for newly mined products. Mining may already be one of the most heavily regulated industries, but the almost inevitable consequence of these trends is that the level of regulation will increase further in the future.

While the effects of this will inevitably feed through to the operating costs of mines, bigger impacts are likely to be felt at the investment stage. Tighter permitting and higher environmental standards will add to the upfront costs of establishing new mines and stretch out already-long permitting processes. This will be challenging for all companies, but particularly for smaller, less well-funded ones under pressure to get into production quickly to keep their shareholders interested. Of equal importance will be uncertainty over the exact environmental regime under which a project will be obliged to operate. When deciding whether or not to commit to investment of hundreds of millions of dollars, companies need to have confidence that their project's economics will not be undermined by large unanticipated changes to the environmental regime under which they are operating. Retrofitting projects to meet new standards is generally more costly than building higher standards into the project at the outset. Uncertainties about the regime governing carbon emissions are a particular concern. Already there is a major debate in the mining industry about 'unburnable carbon', that is, fossil fuels such as coal which are currently considered company assets but which could cease to have value if the imposition of a high charge on carbon generation renders their extraction uneconomic.

These above ground environmental factors collectively represent a growing challenge for the mining industry and for investment in the industry. As a potential barrier to mineral development, they also have implications for mineral availability and, possibly, prices. In the past, the debate about long-term mineral availability has tended to focus on what is in the ground and how to recover minerals cost-effectively. The argument about 'peak' minerals follows in this tradition. It could be, however, that the real limiting factor turns out to lie not what is in the ground but in the capacity of the planet to absorb the environmental side effects of mining and of the goods that mined products are used to produce. For the most part, it would appear that the world has the resources required to meet likely future demand for minerals, the technologies to recover the needed minerals and the companies and financial institutions to organise and fund the mining. What is less clear is where are the limits of society's tolerance for the environmental impacts of the increase in mineral production that this effort will entail and the price that society will demand for mitigating these impacts.

Above ground issues – political

The other critical above ground issue facing the industry is political uncertainty.

Chapter 7 discussed how rising commodity prices during the boom boosted expectations among mineral-rich countries about the potential of mining to provide a springboard to economic development. This was often associated with a view that in order to extract the full value from the sector and ensure that it developed in accordance with the national interest, governments needed to adopt a more interventionist approach. The case was made in the same chapter that this shift in perspective has structural as well as cyclical underpinnings and is unlikely to reverse soon. The politicisation of mining, which this shift has brought with it, is often at odds with the law-based approach to mine development which Western companies in particular want and like, and leaves them more vulnerable to capricious change in the terms and conditions under which they operate.

It has been argued that a further reason for supposing that the resurgence of nationalist attitudes in the mining sector is not likely to dissipate quickly is that it taps into a tendency towards economic and political nationalism which is much wider than the mining sector. In the years following the collapse of the Soviet Union, the world experienced a period during which the United States was left as the only global superpower and in which liberal democratic values were in the ascendancy. This was what might be described as the golden age of globalisation and of international rules-based systems of trading and investment. The emergence of China as a global economic power in the 2000s challenged this so-called Washington consensus, and the GFC effectively killed it off. Since that time, it has become commonplace to refer to a 'multi-polar' world or a 'G0' (G-Zero) world – that is, a world without a dominant superpower.[21] This is an increasingly fragmented world, a world in which economic and political power are more dispersed than at any time since the latter part of the seventeenth century, a period during which the nationalist philosophy of mercantilism prevailed.

While the period following the GFC did not result in a massive upsurge of protectionism as had happened following the Great Crash of 1929, it nevertheless resulted in creeping nationalism pretty much everywhere – a sort of globalisation in reverse. The critics of globalisation and multilateralism have become progressively more vociferous, and a philosophy of 'each country for itself' has spread. Companies which not so very long ago were boasting about how off-shoring was enabling them to reduce their costs, are now lauding the benefits of bringing their business back home, or 're-shoring'. The Doha Round on the trade liberalisation has all but collapsed. The major banks, which helped fund the spread of globalisation, have been massively weakened by the GFC. Their reduced resources and capacity to manage risk means that capital is not flowing across national borders with the same alacrity with which it once did. UNCTAD's *World Investment Report* shows nominal inward FDI flows declining from over $2000 billion in 2007 to $1330 billion in 2012. Takeovers of domestic assets by foreign companies are opposed on what are claimed to be objective grounds of 'national interest' but which in reality are no more than an appeal to populist sentiment. The resource nationalism which is part of this tendency is not ideological in the sense that it is driven by a dislike of capitalism. It is rooted in political pragmatism and a populist distrust of foreigners.

The reassertion of minerals as a sovereign issue is well illustrated by the case of Indonesia, discussed in Chapter 7. Whatever the economic rationale for Indonesia's measures requiring increased domestic ownership of the mining industry and its insistence that minerals be processed before export, and however much the latter policy may run counter to the trade rules of the WTO, the fact that a change of government in 2014 did not result in any change to these policies suggests a fairly high degree of political support for the measures in the country. There have undoubtedly been short-term costs to Indonesia's economy resulting from lost export revenues and mining taxes and a loss of trust among long-term Western investors who have seen their agreements with the government disregarded, but there seems little doubt that the effects of the policies will eventually result in Indonesia growing its role as a smelter and refiner of nickel, copper and alumina, as well as making it a bigger generator of electricity from its local coal resources. China has been the major provider of capital for these developments, often in association with Indonesian partners.

The tendency to economic nationalism, while often associated with the growing role of the state in economic affairs, is not necessarily indicative of the growing strength of government. Indeed, hostility towards foreigners is often used by governments to distract attention from political weakness. The GFC and the subsequent slowing of the global economy undermined the respect for, and authority of, governments throughout the world. While things were booming, there was something for everyone. The crisis and its aftermath revealed the incompetence, inefficiency and corruption of many governments. The governments of advanced economies have surrendered much public support as central banks have printed money to rescue a situation they played a large part in creating. Where governments have sought to rein back spending, they have often brought people out onto the streets. China, where it has long been apparent that the authority of the China Communist Party (CCP) leadership rests on its ability to deliver growth, has seen an increased incidence of public protest. In the vacuum created by the loss of central authority, interest groups – be these the poor, ethnic minorities, regional groupings or local communities – have been empowered. Attempts to roll this back through a reinforcement of central authority will be hard and could be explosive.

The lost authority of government and the tendency of countries to turn inwards puts foreign investors in an invidious position, more so when the investors in question are focused on such quintessential national assets as mineral resources. Weak government can too readily invoke the national interest as justification for bringing pressure to bear on companies and demanding new concessions from them. Political parties vying for office can use their proposed toughness towards the mining industry to garner votes. Local communities and ethnic groupings can exploit their capacity to prevent mining as a means to extract concessions from central government on wholly other agendas, be these welfare, distribution of revenues or indigenous peoples' rights. Mining becomes, in effect, a proxy for other, unrelated agendas. Research by the Royal Institute of International Affairs shows disputes in the extractives sector escalating dramatically in the period since 2003.[22] However high their standards, there is only so much that mining companies can do to protect themselves in these situations. They are up against forces bigger than the industry. In some instances, their only option may be to walk. The deals they do can only be as good as the counterparties to those deals.

How far this tendency to economic and political nationalism, combined with the loss of government authority, will evolve is hard to know. At the time of writing, it would appear still to have some way to run. Self-evidently it has important implications for the flow of investment in mining. Political risk is notoriously difficult for companies, or indeed anyone, to evaluate. It may be that weaker commodity prices will inject a greater degree of realism into mining countries and result in them backing off from some of their more robust nationalist policies. In all probability, there will be some rechanneling of investment flows towards countries where miners feel more secure, even where these countries do not have the best resources. This will contribute to inefficiencies in the allocation of funds for resource development and potentially lead consumers to have to pay more for their minerals than might otherwise have been the case.

Industry directions

The situation in which the mining industry finds itself is not one it expected a decade ago. The commodity boom has passed, and the forces of globalisation are in retreat. The industry's task is to grasp the realities of the world into which it is moving and configure itself so as to be able to operate effectively within it.

Some of the thinking that shaped industry strategies during the commodity boom has been called into question. The notion that in a globalising economy mining companies needed to be ever bigger took a body blow when the attempt to push it to its logical conclusion in 2008 and 2009, most notably with the attempt of Vale to acquire Xstrata and BHP Billiton to acquire Rio Tinto, failed. It also became clear, for example, from the experiences of Rio Tinto in Mongolia and Guinea, that a company's size offers only limited protection from political pressures. In a slower-growth environment, investors too were beginning to question the virtues of gigantism. Size tends to bring with it a loss of focus and bloated head offices. Moreover, large companies in a relatively small industry are inherently vulnerable to regulatory (anti-competition) restrictions and are constrained in their scope for further growth, turning them in investment terms into something close to utilities.

Another notion which was punctured was that improvements in companies' environmental and social performance, and greater transparency in reporting, would make them attractive as investors to mineral host nations. While there can be little doubt that the standards of the industry increased significantly, the hoped-for benefits in terms of enhanced access were in large part negated by the increased politicisation of government attitudes towards the industry.

The politicisation of the industry means that global miners are disadvantaged in certain countries, for example, those of the Former Soviet Union and China, alongside national mining champions. Not only are local companies likely to be better positioned to know about, and to develop, resources available locally, but they can usually ensure that they acquire stakes in any interesting resources unearthed by foreign explorers. In their activities overseas, they can draw on the support of their governments to provide political backing, including sometimes support in the form of development aid for associated infrastructure projects. Increased state involvement in the industry also changes the risk calculus facing investors. Companies with state backing, including sovereign wealth funds, who view resource developments as strategic, can generally be expected to have a different capacity for risk relative to companies who look to stock markets for their financing. More problematic still, the politicisation of decision-making on resource development creates opportunities for private institutions to do deals in a manner not available to those with stock market listings who are obliged to conduct their affairs under the spotlight of public scrutiny. Attempts by regulatory authorities in the United States and European Union (EU) to force publicly quoted companies to meet ever higher standards of behaviour and disclosure, although doubtless well intentioned, risk having the perverse effect of driving miners away from public listings and into the arms of private concerns operating out of public view. The reversion of Eurasian Natural Resources Corporation (ENRC) to a private concern is a case in point. The growth of private equity companies in mining is another.

The changed conditions facing the industry have naturally prompted a strategic response from the GDMs. Weaker commodity prices coupled with the need to pay down debt and to appease disenchanted investors has led the industry to a renewed focus on cost cutting and capital conservation. There has been a trimming of global ambitions and a renewed emphasis on tried and trusted geographies. BHP Billiton has taken to emphasising to investors its status as an OECD miner. Many projects have been cancelled or deferred. BHP Billiton, in 2012, announced the shelving of no less that $40 billion worth of projects in copper, coal and iron ore. There have also been the beginnings of industrial restructuring – 'right sizing' in the jargon – as companies have sought to attain a blend of scale and flexibility more suited to the times. Companies have begun to shed some of their lesser quality or smaller businesses to enable them to concentrate their firepower on their premium assets and put themselves in a position where they can offer investors some growth again.

Rio Tinto sold off some of its 'non-core' assets in coal, talc and aluminium fabrication in order to pay down its debt in the aftermath of the Alcan acquisition. It tried, although without success, to sell its Pacific Aluminium business. BHP Billiton announced in August 2014 that it would be divesting its aluminium, manganese, and nickel businesses, to all intents and purposes unwinding the incorporation of Billiton into BHP which took place in 2001 and the acquisition of WMC Resources in 2005. The divested assets were to go into a new company called South32. Vale said it was looking at spinning out its base metal assets, which include those it acquired with Inco.[23] Anglo American has divested its construction materials business and put some of its Latin American copper assets up for sale. There has been talk of spinning out its platinum and diamond interests or possibly creating a separate company for its South African assets. Again, this has some of the hallmarks of the unwinding of an earlier amalgamation, that with Minorco in 1999.

Running counter to this tendency to corporate dismemberment was the merger between Glencore and Xstrata, announced in 2012 and completed in 2013. Glencore (as the combined entity was subsequently named) was not, however, a GDM in the traditional mould. In keeping with the times and with Glencore's business origins, the company embraces a trading philosophy with a tight focus on operating margins and a stated aversion to large capital projects. Alone among the GDMs, it appeared to have ambitions and a mandate from shareholders to grow through mergers and acquisitions (M&A), in the summer of 2014, reportedly proposing a merger to the board of Rio Tinto.[24]

Such a process of divestment, re-sizing and refocusing resolves some of the problems facing the large companies, although it may create others. The singular success of iron ore had led to these companies becoming dependent on a narrower range of commodities than had historically been the case. As of 2013, one commodity, iron ore, accounted for over 90 per cent of Vale's net profits, over 80 per cent of Rio Tinto's, 50 per cent of BHP Billiton's and 40 per cent of Anglo American's. The divesting of non-core assets by the GDMs clearly sacrifices some of the commodity diversity which has traditionally been deemed one of their key strengths. The divestments do, however, create an opportunity for the rebuilding of the middle ground of mining – an area largely laid waste by the M&A boom of the 2000s. A number of companies with a small clutch of mines (say, between one and five) came through the GFC in relatively good shape, leaving them well positioned to grow themselves through organic growth or merger. There are also several private equity funds looking to move into and repopulate this middle ground, including one, X2 Resources, headed by Mick Davis, formerly of Xstrata. As of the second half of 2014, these funds were believed to have up to $20 billion available to spend on acquisitions.

China's companies will likely be a growing force in global mining, albeit that they may take on a more commercial and less strategic aspect. They are still likely to benefit from cheaper finance and strong backing from the state and from Chinese construction companies, but in other respects their behaviour will fall more closely into line with those of global miners from elsewhere. Mine production in China, on the other hand, may well contract as the industry consolidates and becomes subject to tighter regulation and higher environmental standards. China is overproducing from its limited resource base and, over time, like Japan, will benefit economically from drawing more of its raw materials from better-quality resources elsewhere.

By contrast, mine production from Africa should by rights rise over the next 30 years. The continent has the resources, the people and the development needs. What it has lacked has been governance and infrastructure. No less than 30 African countries routinely suffer from power shortages. In principle, these are resolvable issues, and they are ones where the international community could do more to assist. The cost of redressing Africa's infrastructure deficit has been estimated at

$38 billion a year.[25] The recent commodity boom saw pockets of success in Africa, for example gold and iron ore in West Africa and copper on the copper belt of Zambia and DR Congo, but, for the most part, the opportunity was missed, and Africa still falls well short of fulfilling its potential as a global mine producer. The retreat of commodity markets since 2011 is not helpful, but future cyclical upswings will bring further opportunities.

Licence to mine

A recurring theme of this book is that the mining industry is shaped by developments in the broader economy and in the geopolitical context within which it operates.

The commodity boom of the 2000s put mining and minerals back on the investor and policy agenda and made resource use and resource availability once more a matter of public discourse. It served to demonstrate to governments and the public the importance of a reliable supply of minerals in enabling people the world over to meet their development aspirations and to fashion new products. The growing demand from emerging economies as they seek to improve their lot in coming years will keep minerals in the public eye, even as the commodity cycle swings back and forth.

But if it was the demand for minerals which reaffirmed their economic relevance and which will shape future price cycles, it is on the supply side that much of the drama of mineral markets takes place; where miners battle to identify and prove up resources; where they strive to interest investors in what they are doing and raise the necessary finance; where they grapple with the vagaries of mineral markets and stubborn costs; and where they seek to address the evolving demands of host governments and host communities. These were always challenging matters and they will not get easier. As an industry which involves bringing wealth out of the ground – often someone else's – mining naturally evokes strong emotions and is always susceptible to shifting geopolitical currents.

The commodities boom of the 2000s furnished huge opportunities for mining companies and led to an extensive restructuring and reorientation – or, as termed here, 'remaking' – of the industry as it sought to grasp these opportunities. It helped to restore a lost sense of adventure among mineral explorers, and it drew into the industry a new generation of employees. It also unleashed some powerful forces which will continue to reverberate for many years to come. It resulted in certain mineral markets becoming less structured; it brought to the fore countries and companies which previously played a only a marginal role in mineral markets; and it triggered a wave of resource nationalism across the world and increased state intervention in the affairs of the industry.

The large mining companies are powerful agents for the supply of mineral raw materials. They harness enormous amounts of technical, professional and financial skills. They have the continuity and long-term view which allows them to work on a time scale relevant to the construction and operation of very large capital projects. Their span and visibility gives them a strong vested interest in behaving in a socially responsible manner. However, they also face some major challenges. These include increased uncertainty in the direction of long-run prices and costs, growing regulatory pressures resulting from mining's environmental impacts and, quite possibly (in view of the weakened global banking system), increased funding challenges. They also face a world which is fragmenting politically and which appears increasingly susceptible to populism.

There is doubtless more that mining companies can do to raise their standards of performance and secure the trust of mineral host countries. But to meet their many challenges and fulfil their function, mining companies more than ever need the constructive support of governments. It is only government which has the authority to create the legal framework and stable conditions which large decade-long investment projects require. Yet it is an unavoidable conclusion of the past few years that many governments do not understand the industry well. Governments of many advanced economies have bemoaned the threat to the supply of key minerals but have been reluctant to do anything to encourage the production of those self-same minerals, whether at home or overseas. Governments of many mineral-rich countries have, understandably, sought to ensure that their economies extract the full economic benefits from mining, but their policies to achieve this have often demonstrated a lack of awareness regarding what investors require to induce them to commit to large-scale capital-intensive projects in a cyclical industry. It has been argued above that the world does not have a resource problem. It cannot, however, be ruled out that it may face an investment problem.

In the introduction, reference was made to the fact that during the boom years, the idea emerged in some quarters that mining is an easy business in which to make money. It is not. It may have its moments when this appears to be the case, but there are longer periods in between when it is a margin business and profits have to be ground out in much the same way that minerals have to be ground out from the barren rocks that contain them. The considerable economic and commercial challenges facing the mining industry clearly have the potential to drive the industry to become unduly risk averse, as was the experience in the 1990s, something that is in the interests of neither investors nor consumers. Suitably empowered, the market, and those who seek to serve it, should be more than capable of furnishing the mineral supplies the world of the future will need. Whether they will be allowed to do so will have to be seen.

Notes

Introduction

1. S.D. Strauss (1986) *Trouble in the Third Kingdom* (Mining Journal Books, London).
2. ICMM (2012) *Trends in the Mining and Metals Industry*, International Council on Mining and Metals.
3. The World Bank (2008) *Working Together: How Large-Scale Mining Can Engage with Artisanal and Small-Scale Miners*. Available at: www.icmm.com/document/789
4. N. Shemer, 'Ranking the World's Biggest', *Mining Journal*, 27 June 2014.
5. For example, P.C.F Crowson (2008) *Mining Unearthed* (Aspermont, London); M. Radetzki's (2008) A *Handbook of Primary Commodities in the Global Economy* (Cambridge); chapters by P.C.F. Crowson and D. Humphreys in P. Darling (ed.) *SME Mining Engineering Handbook*, 3rd edn. (Society for Mining, Metallurgy, and Exploration, 2011) and J.E. Tilton (2003) *On Borrowed Time?* (Resources for the Future, Washington DC).

1 The Barren Years

1. H.M. Morgan, Address to the LME annual dinner, Grosvenor House, London, 12 October 1999.
2. A. Maddison (2006) *The World Economy*, OECD, p.613.
3. A. Maddison (2006) *The World Economy*, OECD, p.613.
4. World Bank, *World Bank Indicators,* at http://data.worldbank.org/indicator
5. This distinction is found in J.E. Tilton (1986) 'Beyond Intensity of Use', *Materials & Society*, 10, 3.
6. D. Coyle (1997) *The Weightless World* (Capstone), pp.viii, xii, xiii.
7. D. Humphreys, 'New Economy? We're Living in the Material World', *The Independent on Sunday*, 9 April 2000.
8. W. Leontief, J.C.M. Koo, S. Nasar and I. Sohn (1983) *The Future of Nonfuel Minerals in the US and World Economy* (Lexington Books).
9. IEA and US Bureau of Mines projections are from W. Leontief *et al., The Future of Nonfuel Minerals in the US and World Economy*. Data for 1950–1980 are from Metallgesellschaft, *Metal Statistics*, various editions.
10. US Bureau of Mines, *Minerals Year Book*, 1974 and 1980 editions.
11. Industry sources.
12. S.D. Strauss (1986) *Trouble in the Third Kingdom* (Mining Journal Books), p.3.
13. K. Takeuchi (1990) 'Japan's Experience in Linking Foreign Direct Investment and Imports of Minerals', *Resources Policy*, 16, 4.
14. Metals Economics Group (1992), *Corporate Exploration Strategies*, 3rd edn.
15. J. McNish (1998) *The Big Score* (Doubleday Canada), pp.320–322.

16. J. Otto, 'Fiscal Decentralization and Mining Taxation', *Report prepared for The World Bank Group Mining Department*, March 2001, available at www.revenuewatch.org
17. G. Bridge (2004) 'Mapping the Bonanza: Geographies of Mining Investment in an Era of Neoliberal Reform', *The Professional Geographer*, 56, 3.
18. D. Kosich, 'Guyana High Court Dismisses $2B Omai Gold Mines Tailings Accident Suit', *Mineweb*, 1 November 2006, available at www.mineweb.com
19. D. Humphreys (2013) 'Minerals: Industry History and Fault Lines of Conflict' in R. Dannreuther and W. Ostrowski (eds) *Global Resources: Conflict and Cooperation* (Palgrave Macmillan).
20. J.D. Sachs and A.M. Warner (1995) 'Natural Resource Abundance and Economic Growth', *National Bureau of Economics Research Working Paper 5398*.
21. C.O. Holliday Jr, S. Schmidheiny and P. Watts (2002), *Walking the Talk: The Business Case for Sustainable Development* (Greenleaf Publishing).
22. R. Wilson, Report on the Governors Meeting for the Mining and Metals Industry, 31 January–1 February 2000, Davos, Switzerland.
23. R. Wilson, Report on the Governors Meeting for the Mining and Metals Industry.

2 China Changes Everything

1. M. Farooki and R. Kaplinsky (2012) *The Impact of China on Global Commodity Prices* (Routledge), pp.15–16.
2. World Bank, *World Development Indicators*, at http://data.worldbank.org
3. V. Tulpulé, 'Economic Outlook and Commodity Prices', 29 June 2012, available at www.riotinto.com
4. M. Komesaroff, 'Travels in the Middle Kingdom', *Presentation for UBS*, 7 June 2012.
5. These figures were originally taken from the World Bank's *World Development Report*, possibly for 2003. A similar tabulation is included in McKinsey Global Institute's *Urban World: Cities and the Rise of the Consuming Class*, June 2012, p.4.
6. BP (2014) *Statistical Review of World Energy*, available at www.bp.com/statisticalreview
7. International Monetary Fund (2014) *World Economic Outlook Database*, available at www.imf.org
8. D. Wilson and R. Purushothaman (2003)'Dreaming with BRICS: The Path to 2050', *Goldman Sachs Global Economics Paper No: 99*.
9. Various sources, but particularly J. Anderlini 'A Workshop on the Wane' and G. Magnus, 'China Needs a Long-Term Solution to Its Property Woes', both *FT.Com*, 16 October 2011.
10. 'Reflating the Dragon', *The Economist*, 15 November 2008.
11. Bloomberg, 'China SRB Bought 235,000 tons of Copper This Year', *Bloomberg*, 9 November 2009, available at www.chinamining.org
12. D. Alpert (2013) *The Age of Oversupply* (Portfolio, Penguin), pp.81–87.
13. International Comparison Programme (2014) *Purchasing Power Parities and Real Expenditures World Economies*, at http://icp.worldbank.org/

14. National Bureau of Statistics, 'The Contribution of Three Major Demand for GDP Growth in 2009', *NBS News Release,* 3 February 2010.
15. D. Stanway, 'China's Steel Sector Troubles Can't Be Solved Easily – CISA', *Inside Metals, Thomson Reuters*, 25 February 2014.
16. Author's calculations based on data from NBS.
17. J. Anderlini, 'China Has 'Wasted' $6.8tn in Investment, Warn Beijing Researchers', *FT.Com*, 27 November 2014.
18. L. Lewis, 'Cracks Widen in Beijing's Grand Vision for the Future', *The Times,* 30 August 2011.
19. 'Xi Who Must Be Obeyed', *The Economist*, 20 September 2014.
20. Author's calculations based on data for 2012 from IMF (April 2014 WEO Database), WBMS and WSA.
21. 'The East Is Grey', *The Economist*, 10 August 2013.
22. Data from M. Komesaroff, quoted in E. Reguly (2011) 'China Wants to Drink Our Milkshake', *The Globe and Mail*, 26 May 2011 and in T. Holland (2011) 'Financial Gap Threatens to Sink Beijing's Ambitious Water Goal', *South China Morning Post*, 1 April 2011. Also, 'All Dried Up', *The Economist*, 12 October 2013.
23. Quoted in M. Wolf, 'How China Could Yet Fail Like Japan', *FT.Com*, 14 June 2011.
24. 'The Xi Manifesto', *The Economist*, 23 November 2013.
25. Michael Pettis, 'China Must Bridge the Growth Gap', *FT.Com*, 14 March 2011.
26. G. Magnus, 'This Time China's Property Bubble Really Could Burst', *FT.Com*, 16 May 2014.
27. For example, M. Hovers, 'Iron Ore Market Outlook', *BHP Billiton Presentation*, 6 October 2014, available at www.bhpbilliton.com

3 Feeding the Dragon

1. 'A Great Wall of Waste', *The Economist*, 21 August 2004.
2. FAI data from National Bureau of Statistics, available at www.stats.gov.cn/english
3. This section draws on D. Humphreys 'The Elephant in the Mine', *Mining Journal*, 29 August 2014.
4. *China Metallurgical Newsletter*, 28 February 2013.
5. National Bureau of Statistics (2013), *China Yearbook 2013*. Data relate to 2012.
6. Y. Yang, 'Coal Mining and Environmental Health in China', 2 April 2007, available at www.wilsoncenter.org
7. World Gold Council (2014) *China's Gold Market: Progress and Prospects*, p. 82.
8. Macquarie Capital, *Commodities Comment*, 14 July 2014.
9. 'Looking Inwards', *Mining Journal*, 24 September 2010. Also, 'The Pendulum Swings Against the Pit', *The Economist*, 17 October 2009.
10. 'China Reins in Foreign Investment in Mining Sector', *China Metals*, 10 December 2007.
11. CRE, 'China Halts Hundreds of Mining Projects', *China Rare Earths Information Net*, 20 February 2012, available at www.brire.com/cre/english

12. US Geological Survey (2014), *Mineral Commodity Summaries 2014*, and US Energy Information Administration (EIA) website, available at www.eia.gov
13. Quoted in D. Moyo (2012) *Winner Takes All*, Penguin Books, p. 27.
14. 'China to Consolidate Mines', *China Metals*, 25 January 2007.
15. W. Gu, 'Myths Around China's Revitalisation Plan', *Reuters News*, 25 March 2009.
16. C.-W. Yap, 'China Plans to Create Iron-Ore Mining Giant', *Wall Street Journal*, 20 March 2014, available at http://online.wsj.com
17. 'Chinese Tax Changes Take Market by Surprise', *Metal Bulletin*, 6 November 2006.
18. Cerium is almost as abundant in the earth's crust as copper. Lanthanum is more abundant that lead, tin, cobalt or the platinum group metals (PGMs).
19. Roskill Information Services (2011) *Rare Earths & Yttrium: Market Outlook to 2015*, 14th edn.
20. 'Situation and Policies of China's Rare Earth Industry', at http://news.xinhuanet.com/english/business/2012-06/20/c_131665123.htm
21. 'Mutual Convenience', *The Economist*, 15 March 2008 and M. Komesaroff, 'China Eyes Congo's Treasures', *Far Eastern Economic Review*, March 2008.
22. 'Copper Reserves at China's Sicomines in Congo Less than Hoped', *Inside Metals, Thompson Reuters*, 24 May 2013.
23. This point and several others addressed in this section Are covered in D. Humphreys, 'China Goes Out', *Mining Journal*, 30 November 2012, and D. Humphreys (2012), 'Transatlantic Mining Corporations in the Age of Resource Nationalism', *Transatlantic Academy Paper Series*, Washington, DC.
24. B.F. Tabi Tabe, 'China Steps up Iron Ore Drive in Africa', *Inside Metals, Thompson Reuters*, 18 August 2011.
25. J.D. Wilson (2012) 'Chinese Resource Security Policies and the Restructuring of the Asia-Pacific Iron Ore Market', *Resources Policy*, 37, 3.
26. A. Schipani, 'Peru to Become World's Second-Largest Copper Producer', *FT.Com*, 25 August 2014.
27. A. Desir, 'Changing Lanes', *Mining Journal*, 7 June 2013.
28. 'More than Minerals', *The Economist*, 23 March 2013.
29. 'Sierra Leone and China Sign Deal Worth US$8bn', *Mining Journal*, 12 July 2013.
30. L. Sanusi, 'Africa Must Get Real about Chinese Ties, *FT.Com*, 11 March 2013.
31. M. Komesaroff, 'Screwing up in Foreign Climes', *China Economic Quarterly*, June 2012.
32. L. Hornby, 'Anchors Finally Away for Chinese Australian Mine, *FT.Com*, 4 December 2013 and H. Sender 'China: Dug in too Deep', *FT.Com*, 24 June 2012.
33. A. Desir, 'Betting on Success', *Mining Journal*, 14 March 2014.
34. Minmetals, for example, abandoned its attempt to acquire Zambian copper miner Equinox in April 2011 when faced with a superior bid from Barrick Gold.
35. E. Lin, 'China's Great Leap into Africa', *Mining Journal Special Publication – Indaba*, January 2013.

36. D. Moyo (2012) *Winner Takes All.*
37. M. Ericsson, 'China FDI in Mining: Threat or Opportunity?' *Commodities Now: LME Week Supplement,* October 2011.
38. A. Desir, 'Seeing Red: The Truth Behind China's Outbound Investment', *Mining Journal,* 8 June 2012 and A. Desir, 'Betting on Success', *Mining Journal,* 14 March 2014.

4 Prices Take Wing

1. For a fuller discussion on these issues, see D. Humphreys (2011) 'Pricing and Trading in Metals and Minerals' in P. Darling (ed.) *SME Mining Engineering Handbook,* 3rd edn. (Society for Mining, Metallurgy, and Exploration).
2. This and the following paragraphs draw heavily on D. Humphreys (2010) 'The Great Metals Boom: A Retrospective', *Resources Policy,* 35, 1.
3. A. Heap, 'Riding the Super Cycle', *Citigroup Global Equity Research,* 31 January 2005.
4. J.T. Cuddington and D. Jerrett (2008) 'Super Cycles in Real Metals Prices?', *IMF Staff Papers,* 55, 4.
5. For example, K. Cooper, 'Are Gold and Oil Due a 20-year Boom?' *The Sunday Times,* 19 June 2005.
6. M. Radetzki, R.G. Eggert, G. Lagos, M. Lima and J.E. Tilton (2008) 'The Boom in Mineral Markets: How Long Might it Last?', *Resources Policy,* 33, 3. Also, M. Stürmer, '150 Years of Boom and Bust – What Drives Mineral Commodity Prices?', *German Development Institute Discussion Paper 5/2013.*
7. J. Blas (2012) 'A Less-Super Commodities Supercycle', *FT.Com,* 29 May 2012. J. Jessop, 'The Not-So Super-Cycle', *Commodities Now,* June 2012. R. Sharma, 'We Should Celebrate the End of the Commodity Supercycle', *FT.Com,* 24 June 2012.
8. H.R. Jansen *et al.*, 'Metals & Mining: Super Cycle Sunset', *Citi Global Equities Research,* 21 March 2012.
9. J.E. Tilton (2014) 'Cyclical and Secular Determinants of Productivity in the Copper, Aluminium, Iron Ore, and Coal Industries', *Mineral Economics,* 27, 1.
10. D. Alpert (2013) *The Age of Oversupply* (Portfolio, Penguin). Also, R. Freeman, 'China, India and the Doubling of the Global Labour Force', *The Globalist,* 3 June 2005.
11. Quoted in D. Humphreys, 'The Great Metals Boom: A Retrospective'.
12. M. Wolf, 'The World Must Learn to Live with a Wide-Awake China', *FT.Com,* 12 November 2003.
13. J.T. Cuddington and D. Jerrett (2008) 'Super Cycles in Real Metals Prices?'
14. D. Humphreys (2012) 'Supercycle Wobbles', *Mining Journal,* 16 March 2012.
15. D. Kosich, 'Proponents Seek CFTC Speculative Trading Caps to Protect Metals Consumers and Metals Markets', *Mineweb,* 26 March 2010.
16. J. Rogers (2007) *Hot Commodities* (Chichester, West Sussex, England, John Wiley).
17. J. Grantham, 'Time to Wake Up: Days of Abundant Resources and Falling Prices are Over Forever', *GMO Quarterly Letter,* April 2011.

18. The websites of the respective funds at http://us.spindices.com and www.djindexes.com
19. G. Gorton and K.G. Rouwenhorst, 'Facts and Fantasies about Commodities Futures', *Yale ICF Working Paper No 04–20*, 14 June 2004 (28 February 2005 draft).
20. US Commodity Futures Trading Commission, available at www.cftc.gov/MarketReports
21. Thomson Reuters GFMS, *Gold Survey 2013*.
22. M.W. Masters, 'Testimony before the Commodities Futures Trading Commission', 5 August 2009.
23. Quoted in J. Kemp, 'Controversial Research Behind Commodity Rules', *Reuters*, 11 January 2010.
24. These matters are discussed in detail in J.E. Tilton, D. Humphreys and M. Radetzki (2011) 'Investor Demand and Spot Commodity Prices', *Resources Policy*, 36, 3. See also, O. Östensson (2012) 'The 2008 Commodity Price Boom: Did Speculation Play a Role?' *Mineral Economics*, 25, 1.
25. UNCTAD (2009) 'The Financialisation of Commodity Markets', Chapter 2 of *Trade and Development Report*, 2009.
26. IMF (2006), *World Economic Outlook* (Washington, DC, International Monetary Fund), Chapter5.
27. HM Treasury (2008) *Global Commodities: A Long Term Vision for Stable, Secure and Sustainable Global Markets* (London: HM Treasury). US Interagency Task Force on Commodity Markets (2008) *Interim Report on Crude Oil* (Washington DC: US Interagency Task Force on Commodity Markets). S.H. Irwin and D.R. Sanders, 'Speculation and Financial Fund Activity, Draft Report, Annex' in *Proceedings of OECD Working Party on Agricultural Policies and Markets,* 17–20 May 2010.
28. IMF (2010) *Global Prospects and Policies* (Washington, DC, International Monetary Fund), Chapter 1, Appendix 1.
29. J. Kemp, 'Commodity Indices: Struggling Against the Tide', *Commodities Now*, September 2011.
30. Website for S&P Dow Jones Indices, available at http://us.spindices.com
31. G. Meyer, 'US Funds Fall Out of Love with Commodities', *FT.Com*, 11 September 2012.
32. CFTC, Index Investment Data, available at www.cftc.gov
33. J. Farchy, G. Meyer and J. Blas, 'Commodity Hedge Fund Investors Seek Exit', *FT.Com*, 6 February 2013, and C. Milhench, 'Commodity Funds Face Uphill Struggle after Dire 2013', *Inside Metals, Thomson Reuters*, 16 January 2014.
34. A.L. Sussman and E. Stephenson, 'Worried Fed Seeks to Curb Wall Street Bank Commodity Trade' *Inside Metals, Thomson Reuters*, 13 January 2014.
35. 'Sell Signals', *The Economist* 26 April 2014.
36. 'Metal Bashing', *The Economist*, 17 August 2013.
37. P. Waldmeir and W. MacNamara, 'Rio Tinto Case Highlights Risks in China', *FT.Com*, 5 April 2010.
38. M. Sainsbury, 'Rio Tinto's Stern Hu Jailed for 10 Years', *The Australian*, 29 March 2010.
39. A. Home, 'Copper Concentrates Pricing Starts to Fragment', *Inside Metals, Thomson Reuters*, 16 December 2013.

5 An Industry Transformed

1. X. Rice, 'A Magic Formula that keeps Share Price Bubbling', *The Times*, 19 January 2004.
2. International Copper Study Group, available at www.icsg.org
3. BHP Billiton, Interim Results briefing, 19 February 2004, available at www.bhpbilliton.com
4. Rio Tinto, Interim Results presentation, 29 July 2004, available at www.investegate.co.uk
5. This section draws on D. Humphreys (2012) 'Transatlantic Mining Companies in the Age of Resource Nationalism', *Transatlantic Academy Paper Series*, Washington, DC, pp.3–4.
6. How Xstrata Won Falconbridge, *National Post*, undated (likely late 2006), available at www.financialpost.com
7. M.A. McCrae, 'The Worst Mining Deal Ever: Rio Tinto Buying Alcan for $38.1B', *Mining.Com*, 15 February 2013, available at www.mining.com
8. PwC (2010), *Mine – When the Going Gets Tough…*, Pricewaterhouse Coopers.
9. This section draws on D. Humphreys, 'The Rise of the Global Diversified Miner', *Mining Journal*, 19 March 2010.
10. The idea of globalisation spawned an extensive literature. One of the best contemporary books on the subject was T.L. Friedman (2006) *The World is Flat* (Penguin Books).
11. O. Kreuzer and M.A. Etheridge, 'Risk and Uncertainty in Mineral Exploration: Implications for Valuing Mineral Exploration Properties', *AIG News*, 100, May 2010, available at www.aig.org.au
12. For example, R. Schodde, 'Gold Exploration: Gloom or Boom?' *Special Publication on Gold, Mining Journal*, September 2013.
13. PwC (2008) *Mine: As Good as It Gets?* PricewaterhouseCoopers.
14. J. Bowker, 'Rio Tinto See Metals Demand Doubling in 15 Years', *Reuters News*, 4 February 2010.
15. Capex data are author's estimates, based on a variety of different sources and deflated using the US GDP deflator. Note that these data relate to quoted metal-mining companies and cannot be considered comprehensive. Using a broader definition of the mining industry, it has been estimated that industry capex at the peak of the cycle in 2012 was nearer $200 billion, that is, $75 billion above the total shown in Figure 5.3. See Raw Materials Group, 'Mining Capex Towards 2020 – A Silver Lining', LME Week Supplement, *Commodities Now*, October 2013.
16. D. Humphreys (2014) 'The Mining Industry and the Supply of Critical Minerals' in G. Gunn (ed.), *Critical Metals Handbook* (Chichester, West Sussex, UK: American Geophysical Union and Wiley), p.26.
17. International Copper Study Group, available at www.icsg.org. Accessed October 2014.
18. Morgan Stanley research quoted in R. Roberts, 'Reality Bites', *Mining Journal* 26 September 2014.
19. 'Mines & Money: Underperformance Due to Lower Grades – BlackRock', *Mining Journal*, 7 December 2012.
20. H. Thomas, 'Global Miners Stage "Retreat to the Core" ', *FT.Com*, 11 July 2012.

21. PwC (2014) *Mine: Realigning Expectations*, PricewaterhouseCoopers.
22. Evy Hambro quoted in 'Miners Have Lost Investor's Trust, says BlackRock's Evy Hambro', *The Australian*, 18 January 2013.
23. Quoted in D. Fortson, 'Chemistry Prof Out to Keep BHP from Falling into a Hole', *The Sunday Times*, 16 March 2014.
24. J. Wilson, 'Self-Help Digs Miners BHP and Rio Tinto Out of a Hole', *FT.Com*, 18 February 2014.
25. PwC (2014) *Mine: Realigning Expectations*, PricewaterhouseCoopers.
26. N. Wright, 'M&A Activity Jumps YOY in H1 '14', *SNL Metals & Mining newsletter*, 20 August 2014.

6 Enter the Emergers

1. This section, and indeed chapter, draws heavily on D. Humphreys (2011) 'Emerging Miners and their Growing Competiveness', *Mineral Economics*, 24, 1, pp.7–14.
2. P. Andrews-Speed and R. Dannreuther (2011) *China, Oil and Global Politics*, (London and New York: Routledge), pp.86–87.
3. J. Anderlini 'Foreign Losses Spark Accusation of "Mismanagement" at China Fund', *Financial Times*, 19 June 2014.
4. M. Komesaroff (2012) 'China's Foreign Mineral Adventures' *SAFPI Policy Brief No 18*.
5. PwC (2014), *Mine 2014 – Realigning Expectations*, PricewaterhouseCoopers.
6. USGS (2014) *Mineral Commodity Summaries*.
7. D. Roberston, 'More Soviet than City: the Oligarch's Boardroom Coup', *The Times*, 9 June 2011. Also, W. MacNamara and A. Smith, 'Corporate Governance: The Eurasian Equation', *FT.Com*, 23 June 2011.
8. Research by Morgan Stanley reported in R. Wigglesworth, 'National Champions Shine in EM Value Hunt', *FT.Com*, 25 April 2012.
9. E. Downs (2012) 'China Buys into Afghanistan', *SAIS Review*, XXXII, 2, Summer-Fall.
10. 'Valuable Vale', *The Economist*, 25 September 2010.
11. S. Pearson, 'Brazil Raises Pressure on Vale', *FT.Com*, 23 March 2011.
12. 'Norilsk Nickel to Sell Foreign Assets, Double Russia Exploration', *The Moscow Times*, 12 November 2013.
13. S. Mundy, 'Chinese Accused of Zambian Labour Abuses', *FT.Com*, 20 January 2011.
14. J. Leahy, 'Brazil and China Trade Tension Set to Rise', *FT.Com*, 30 January 2011.
15. F. Wong and J. Chaney, 'Stakes and Loans to Shape China's 2011 Mining M&A Menu', *Reuters*, 10 December 2010.
16. M. Costello, 'All-powerful Boards Risk Driving Shareholders into Militant Action', *The Times*, 9 October 2010. U. Galani, 'Big Mining Enters New Era of Risky Asset Deals, *Reuters*, 19 October 2010.
17. D. Robertson, 'Oligarchs are Ordered to Open their Companies to Ordinary Investors', *The Times*, 15 December 2011. J. Gapper, 'The City has Gone too Far for Minority Investors', *FT.Com*, 26 June 2013.

18. S. Antonioli, M. Burton and A. Davies, 'ENRC Could Sell Precious DRC Mines to Repay Costly Debt', *Inside Metals, Thompson Reuters*, 3 February 2014.
19. J. Wilson, 'Rothschild's Bumi Alliance with Bakries Finally Unwound', *FT.Com*, 25 March 2014.

7 Dividing the Spoils

1. P. Cleary (2011) *Too Much Luck* (Collingwood VIC: Black Inc), p.5.
2. BEA (2014), US Bureau of Economic Analysis website, available at www.bea.gov
3. Statistics Canada (2014), Statistics Canada website, available at www.statcan.gc.ca
4. ABS (2014), Australian Bureau of Statistics website, available at http://ausstats.abs.gov.au
5. I. Satchwell, 'Mining's Legacy: Thinking Beyond the Mine', Presentation to SNL conference *Mining on Top*, London 24 June 2014.
6. ICMM (2014) *The Role of Mining in National Economies (2nd edition)* (International Council on Mining and Metals), p.23.
7. IMF (2013), *World Economic Outlook* (Washington DC: International Monetary Fund).
8. The data point is taken from the OECD at http://stats.oecd.org. 'For a full discussion of these matters', see R. Garnaut (2013) *Dog Days: Australia After the Boom* (Collingwood VIC: Redback).
9. MMSD (2002) *Breaking New Ground: Executive Summary*, Mining, Minerals and Sustainable Development (London: Earthscan), p.5.
10. M. Humphreys, J.D. Sachs and J.E. Stiglitz (2007) *Escaping the Resource Curse* (New York: Columbia University Press).
11. P. Collier and A.J. Venables (2011) *Plundered Nation? Successes and Failures in Natural Resources Extraction* (London: Palgrave Macmillan).
12. McKinsey Global Institute (2013) *Reversing the Curse: Maximizing the Potential of Resource-Driven Economies*, available at www.mckinsey.com
13. African Union (2009) *Africa Mining Vision*, Addis Ababa, Ethiopia.
14. PwC (2012) *Corporate Income Taxes, Mining Royalties and Other Mining Taxes: Summary of Rates and Rules of Selected Countries*, PricewaterhouseCoopers., available at www.pwc.co.uk
15. WGC (2013a) *World Gold Council Guidance Note on Non-GAAP Metrics: All-In Sustaining Costs and All-In Costs*, 27 June 2013, available at www.gold.org
16. Deutsche Bank research quoted in G. Parkinson 'Investors "have not received a fair share" of Boom Mining Profits', *The Times*, 17 December 2011.
17. This section draws on D. Humphreys (2012), *Transatlantic Mining Corporations in the Age of Resource Nationalism*, Transatlantic Academy Paper Series, Washington DC.
18. African Union (2009) *Africa Mining Vision*, February 2009, pp.8, 17.
19. PwC (2013), *Mine – A Confidence Crisis: Review of Global Trends in the Mining Industry – 2013*, PricewaterhouseCoopers, available at www.pwc.co.uk, p.48.

20. WTO (2010) *World Trade Report 2010: Trade in Natural Resources,* (Geneva: World Trade Organization).
21. S. Donnan, 'Trade: Into Uncharted Water', *FT.Com,* 24 October 2013.
22. 'The Gated Globe', *The Economist,* 12 October 2013, and associated supplement.
23. 'The Gated Globe', *The Economist* (2013), supplement, 12 October 2013, p.4.
24. C. Cohn, 'Resource Companies Ripping Off Africa: AfDB', *Reuters,* 16 June 2013, available at http://mobile.reuters.com
25. D. Humphreys (2013) 'Minerals: Industry History and Fault Lines of Conflict' in R. Dannreuther and W. Ostrowski (eds) *Global Resources: Conflict and Cooperation* (London: Palgrave Macmillan).
26. 'Special Report on State Capitalism', *The Economist,* and editorial 'The Rise of State Capitalism', 21 January 2012.
27. This section draws on D. Humphreys, 'For Better or Worse', *Mining Journal,* 7 March 2014.
28. D. Humphreys (2013) 'New Mercantilism: A Perspective on How Politics in Shaping World Metal Supply', *Resources Policy,* 38, 3.
29. W. Ostrowski (2013) 'The Political Economy of Global Resources' in R Dannreuther and W. Ostrowski (eds) *Global Resources: Conflict and Cooperation* (London: Palgrave Macmillan), p.104.
30. K. Brooks and T. Wilson, 'Missed Opportunities', *Mining Journal,* 23 August 2013.
31. The World This Week, *The Economist,* 15 June 2013.
32. 'Barrick and Antofagasta Drop Disputed Reko Diq Project', *Mining Journal,* 10 May 2013.
33. Ernst & Young, *Business Risks Facing Mining and Metals,* editions 2008–2009–2014–2015.
34. Oyu Tolgoi (2014), *Factsheet on Economic Contribution,* available at http://ot.mn.en
35. Simandou (2013) *Simandou Economic Impact Report,* May 2013, available at www.riotintosimandou.com
36. ICMM (2011) *Mining: Partnership for Development Toolkit,* available at www.icmm.com
37. ICMM (2012) *Mining's Contribution to Sustainable Development – An Overview,* available at www.icmm.com
38. WGC (2013b) *Responsible Gold Mining and Value Distribution: A Global Assessment of the Economic Value Created and Distributed by Members of the World Gold Council,* available at www.gold.org
39. A point made in M. Solomon (2012) *The Rise of Resource Nationalism* (Cape Town: South African Institute of Mining and Metallurgy).

8 Consumer Concerns

1. NRC (2008) *Minerals, Critical Minerals, and the US Economy: A Summary,* Report of the Committee on Critical Mineral Impacts on the US Economy, Committee on Earth Sciences, National Research Council (The National Academies Press, Washington DC).

2. T.S. Purdum, 'Clinton Unveils Plan to Halt Gold Mine Near Yellowstone', *The New York Times*, 13 August 1996, available at www.nytimes.com
3. T. Treadgold, 'Rare-Earth Lesson for Copper,' *Mining Journal*, 23 March 2012.
4. D. Humphreys (1995) 'Whatever Happened to Security of Supply? Minerals Policy in the Post-Cold War World', *Resources Policy*, 21, 2.
5. Verbal comment by D. Kingsworth, 'Critical Materials Supply Security: Rare Earths', *Presentation at Chatham House*, 24 April 2012.
6. M. Komesaroff, 'How the US Forfeited Its Lead, and a Framework to Win It Back', *American Metal Market*, April 2010.
7. M. Komesaroff, personal communication.
8. OECD (2011) *OECD Due Diligence Guidance for Responsible Supply Chains of Minerals from Conflict-Affected and High-Risk Areas*, OECD Publishing, Paris, available at http://www.oecd-ilibrary.org
9. European Commission (2008) *The Raw Materials Initiative*, COM(2008) 699 final, available at http://eur-lex.europa.eu
10. European Commission (2010) *Critical Raw Materials for the EU*, Report of the Ad-hoc Working Group on Defining Critical Raw Materials, June 2010, available at http://ec.europa.eu
11. M. Watts, 'Strategic Mineral Policy Rises on EU Agenda', *Industrial Minerals*, January 2011. G. Wiesmann, 'German Move to Secure Raw Materials' *FT.Com*, 30 January 2012.
12. European Commission (2014) Report *on Critical Raw Materials for the EU*, Report of the Ad-hoc Working Group on Defining Critical Raw Materials, May 2014, available at http://ec.europa.eu
13. R. Speight, S. Yapp and C. Pacey, 'Transparency Drive', *Mining Journal*, 26 April 2013.
14. US Geological Survey (2014) *Mineral Commodity Summaries*.
15. NRC (2008) *Minerals, Critical Minerals, and the US Economy*, Report of the Committee on Critical Mineral Impacts on the US Economy, Committee on Earth Sciences, National Research Council (Washington DC: The National Academies Press).
16. DOE (2010) *Critical Materials Strategy*, US Department of Energy, December 2010.
17. Details of the institute are available at https://cmi.ameslab.gov/
18. For example, L. Germain and M. Watts, 'Industry Calls for US Critical Minerals Strategy', *Industrial Minerals*, March 2012.
19. J. Kemp, 'Critical Minerals and Mining Reform in the US', *Inside Metals, Thomson Reuters*, 31 January 2014.
20. 'US SEC Sets March Date for New Extractive Rules', *Inside Metals, Thomson Reuters*, 28 May 2014.
21. 'Conflict Minerals Law Starts Working Despite Poor Industry Response' *Inside Metals, Thomson Reuters*, 1 July 2014.
22. C. Oliver, K. Manson and J. Wilson, 'EU Plans Voluntary Rules on Conflict Mineral Imports', *FT.Com*, 5 March 2014.
23. Office of the US Trade Representative, *WTO Case Challenging China's Export Restrictions on Raw Material Input*, 28 October 2009.
24. These observation are drawn largely from WTO (2011) *China – Measures Related to the Exportation of Various Raw Materials*, Report of the Panel,

(WT/DS394/R, WT/DS395/R, WT/DS398/R) 5 July 2011, available at www.wto.org

25. These are respectively, Articles, XI:2(a), Article XX(b) and Article XX(g) of the GATT, 1994.
26. WTO (2014) *China – Measures Related of Rare Earths, Tungsten and Molybdenum*, Report of the Panel (WT/DS431/R, WT/DS432/R, WT/DS433/R) 26 March 2014, available at www.wto.org
27. Roskill, 'What does the WTO Ruling Mean for the Chinese Rare Earth Industry?' *Roskill Briefing Paper*, June 2014, and A. King 'Analysts Shrug Off Rare Earth Trade Ruling', *Chemistry World*, 2 April 2014, available at www.rsc.org
28. D. McGroarty, 'E Pluribus, Unum?' in 'Critical Materials for Green Energy', *Industrial Minerals*, July 2012.
29. European Commission (2010) *Critical Raw Materials for the EU*, Report of the Ad-hoc Working group on Defining Critical Raw Materials, June 2010, p.6.
30. B. Bruijs and H. Sievers (2011) 'Critical Thinking about Critical Minerals', November 2011. *Briefing Paper Based upon Research Undertaken in the Context of the EU 7th Framework Programme POLINARES Project.* Available at www.polinares.eu
31. Roskill, 'Critical Thinking About Critical Raw Materials in the EU', *Roskill Briefing Paper*, June 2014.
32. C. Chan, 'Critical Materials Issues Cut Both Ways', *Mining Journal*, 28 March 2014.
33. See also D. Humphreys (2012) 'Transatlantic Mining Corporations in the Age of Resource Nationalism', *Transatlantic Academy Paper Series*, Washington DC, p.20.
34. M. O'Driscoll, 'Europe Goes Critical', *Industrial Minerals*, July 2010.

9 The Next Thirty Years

1. M. Garrahan, 'Google Founder Backs Plan to Mine Asteroids', *FT.Com*, 24 April 2012.
2. S.D. Strauss (1986) *Trouble in the Third Kingdom* (London : Mining Journal Books), pp.218–219.
3. D. Stanway, 'China Steel Demand Shrinks for First Time in 14 Years as Slowdown Stings', *Reuters*, 25 September 2014.
4. M. Stürmer, '150 Years of Boom and Bust – What Drives Mineral Commodity Prices?', *German Development Institute Discussion Paper*, May 2013.
5. McKinsey Global Institute (2012) *Urban World: Cities and the Rise of the Consuming Class*. Available at: www.mckinsey.com
6. Roskill, 'Rare Earths Industry Evolution at Metal Events' 10th International Rare Earths Conference', November 2014, available at www.roskill.com
7. 'LME Week Loses Some Lustre as Belts Tighten, Focus Shifts East', *Inside Metals*, 20 October 2014.
8. P. Robinson, 'Copper – Returning to Normality?' *Presentation to CRU Copper Conference*, Santiago, 2011.

9. J. Steyn 'Global Gold Sector', *Citi Equities Research*, 9 October 2014.
10. Quoted in N. Hume, 'Price of Steelmaking Ingredient Falls to a Five-year Low', *FT.Com*, 7 October 2014.
11. David Flanagan of Atlas Iron quoted in 'Australian Iron Ore Miners Hang Tough as Prices Fall', *Inside Metals, Thomson Reuters* 13 March 2014.
12. For example, U. Bardi and M. Pagani (2007) 'Peak Minerals', *The Oil Drum: Europe*, 15 October 2007. Available at: http://www.theoildrum.com/node/3086. Also R.A. Kerr (2014) 'The Coming Copper Peak', *Science*, 343.
13. M. Ericsson and P. Soderholm, (2013) 'Mineral Depletion and Peak Production' in R. Dannreuther and W. Ostrowski (eds) *Global Resources: Conflict and Cooperation* (London: Palgrave Macmillan).
14. M. Chender, *What Does the Next Generation of Opportunities Look Like?* SNL Metals Economics Group, Presentation to Mines & Money Conference, London, 2012.
15. T. Albanese, 'Kennecott Utah Copper Site Visit, 16 May 2008'. Available on Rio Tinto website at www.riotinto.com
16. GFMS Thomson Reuters, *Gold Survey*, various editions.
17. R. Schodde 'Gold Exploration ... a Balancing Act', *Mining Journal Gold Supplement*, August 2014.
18. The following paragraphs draw heavily on D. Humphreys (2013) 'Long Run Availability of Mineral Commodities', *Mineral Economics*, 26, 1–2, 1–11. Data in these paragraphs was sourced from T.E. Norgate *et al.* (2007) 'Assessing the Environmental Impact of Metal Production Processes', *Journal of Cleaner Production*, 15, 838–848 and G.M. Mudd (2008) 'Sustainability Reporting and Water Resources: a Preliminary Assessment of Embodied Water and Sustainable Mining', Monash University, Melbourne.
19. C. Hinde, 'Still Counting', *Mining Journal*, 28 October 2011.
20. Ernst & Young (2014), *Business Risks Facing Mining and Metals 2014–2015*, available at www.ey.com
21. I. Bremmer (2012) *Every Nation for Itself: Winners and Loser in a G-Zero World* (New York : Portfolio).
22. P. Stevens *et al.* (2013) *Conflict and Coexistence in the Extractive Industries* (New York: Royal Institute of International Affairs).
23. J.P. Spinetto, 'Vale Says Base Metals Units May Be Worth up to $35 Billion', *Bloomberg.com*, 2 December 2014.
24. J. Wilson and N. Hume 'Rio Turned Down Glencore Merger Idea' *FT.Com*, 7 October 2014.
25. A.M.A. Pedro 'Infrastructure to Support Industrialisation in Africa', Presentation to *Mining on Top: Africa – London Summit*, June 2014.

Selected Bibliography and Data Sources

African Union (2009) *Africa Mining Vision* (Addis Ababa, Ethiopia: African Union).
Ali, S.H. (2009) *Treasures of the Earth* (New Haven and London: Yale University Press).
Alpert, D. (2013) *The Age of Oversupply* (London: Portfolio Penguin).
Andrews-Speed, P. and Dannreuther, R. (2011) *China, Oil and Global Politics* (London and New York: Routledge).
Bardi, U. and Pagani, M. (2007) 'Peak Minerals', *The Oil Drum: Europe*, 15 October 2007.
Bardi, U. (2014) *Extracted* (White River Junction, VT: Chelsea Green).
Bremmer, I. (2012) *Every Nation for Itself: Winners and Loser in a G-Zero World* (New York: Portfolio).
Bridge, G. (2004) 'Mapping the Bonanza: Geographies of Mining Investment in an Era of Neoliberal Reform', *The Professional Geographer*, 56, 3.
Bruijs, B. and Sievers, H. (2011) 'Critical Thinking about Critical Minerals', November 2011. Briefing Paper Based upon Research Undertaken in the Context of the EU 7th Framework Programme POLINARES Project. Available at: www.polinares.eu
Cleary, P. (2011) *Too Much Luck* (Collingwood, VIC: Black Inc).
Collier, P. and Venables, A.J. (2011) *Plundered Nation? Successes and Failures in Natural Resources Extraction* (London: Palgrave Macmillan).
Coyle, D. (1997) *The Weightless World* (Oxford: Capstone).
Coulson, M. (2012) *The History of Mining* (Petersfield: Harriman House).
Crowson, P.C.F. (2008) *Mining Unearthed* (London: Aspermont).
Cuddington, J.T. and Jerrett, D. (2008) 'Super Cycles in Real Metals Prices?', *IMF Staff Papers*, 55, 4.
Darling, P. (ed.) (2011) *SME Mining Engineering Handbook*, 3rd edn. (Englewood, CO: Society for Mining, Metallurgy, and Exploration).
DOE (2010) *Critical Materials Strategy* (US Department of Energy, December 2010).
Downs, E. (2012) 'China Buys into Afghanistan', *SAIS Review*, XXXII, 2, Summer-Fall.
Ericsson, M. and Soderholm, P. (2013) 'Mineral Depletion and Peak Production' in R. Dannreuther and W. Ostrowski (eds) *Global Resources: Conflict and Cooperation* (London: Palgrave Macmillan).
European Commission (2008) *The Raw Materials Initiative*, COM(2008) 699 final, available at http://eur-lex.europa.eu
European Commission (2010) *Critical Raw Materials for the EU* (Report of the Ad-hoc Working Group on Defining Critical Raw Materials), June 2010, available at: http://ec.europa.eu
Farooki, M. and Kaplinsky, R. (2012) *The Impact of China on Global Commodity Prices* (London and New York: Routledge).

Freeman, R. 'China, India and the Doubling of the Global Labour Force', *The Globalist*, 3 June 2005.

Friedman, T.L. (2006) *The World Is Flat* (London: Penguin Books).

Garnaut, R. (2013) *Dog Days: Australia After the Boom* (Collingwood, VIC: Redback).

HM Treasury (2008) *Global Commodities: A Long Term Vision for Stable, Secure and Sustainable Global Markets* (London: HM Treasury).

Holliday Jr., C.O., Schmidheiny, S. and Watts, P. (2002), *Walking the Talk: The Business Case for Sustainable Development* (Sheffield: Greenleaf Publishing).

Humphreys, D. (1995) 'Whatever Happened to Security of Supply? Minerals Policy in the Post-Cold War World', *Resources Policy*, 21, 2.

Humphreys, D. (2010) 'The Great Metals Boom: A Retrospective', *Resources Policy*, 35, 1.

Humphreys, D. (2011) 'Emerging Miners and Their Growing Competiveness', *Mineral Economics*, 24, 1.

Humphreys, D. (2012), 'Transatlantic Mining Corporations in the Age of Resource Nationalism', *Transatlantic Academy Paper Series*, Washington, DC.

Humphreys, D. (2013) 'Minerals: Industry History and Fault Lines of Conflict' in R. Dannreuther and W. Ostrowski (eds) *Global Resources: Conflict and Cooperation* (London: Palgrave Macmillan).

Humphreys, D. (2013) 'Long Run Availability of Mineral Commodities', *Mineral Economics*, 26, 1–2.

Humphreys, D. (2013) 'New Mercantilism: A Perspective on How Politics in Shaping World Metal Supply', *Resources Policy*, 38, 3.

Humphreys, D. (2014) 'The Mining Industry and the Supply of Critical Minerals' in G. Gunn (ed.), *Critical Metals Handbook* (Chichester, West Sussex: American Geophysical Union and Wiley), pp. 20–40.

Humphreys, M., Sachs, J.D. and Stiglitz, J.E. (2007) *Escaping the Resource Curse* (New York: Columbia University Press).

IMF (2010) *Global Prospects and Policies* (Washington, DC: International Monetary Fund), Chapter 1, Appendix 1.

Irwin, S.H. and Sanders, D.R. 'Speculation and Financial Fund Activity, Draft Report, Annex' in *Proceedings of OECD Working Party on Agricultural Policies and Markets,* 17–20 May 2010.

Kerr, R.A. (2014) 'The Coming Copper Peak', *Science*, 343, 722–724.

Komesaroff, M. 'China Eyes Congo's Treasures', *Far Eastern Economic Review*, March 2008.

Komesaroff, M. 'How the US Forfeited Its Lead, and a Framework to Win It Back', *American Metal Market*, April 2010.

Komesaroff, M. 'Screwing up in Foreign Climes', *China Economic Quarterly*, June 2012.

Komesaroff, M. (2012) 'China's Foreign Mineral Adventures' *SAFPI Policy Brief No 18*.

Leontief, W., Koo, J.C.M., Nasar, S. and Sohn, I. (1983) *The Future of Nonfuel Minerals in the US and World Economy* (Lexington, MA and Toronto: Lexington Books).

Maddison, A. (2006) *The World Economy* (Paris: OECD).

McKinsey Global Institute (2009) *Preparing for China's Urban Billion.*Available at: www.mckinsey.com

McKinsey Global Institute (2011) *Resource Revolution: Meeting the World's Energy, Materials, Food, and Water Needs*. Available at: www.mckinsey.com

McKinsey Global Institute (2012) *Urban World: Cities and the Rise of the Consuming Class*. Available at: www.mckinsey.com

McKinsey Global Institute (2013) *Reversing the Curse: Maximizing the Potential of Resource-Driven Economies*. Available at: www.mckinsey.com

McNish, K. (1998) *The Big Score* (Toronto: Doubleday Canada).

MMSD (2002) *Breaking New Ground: Executive Summary*, Mining, Minerals and Sustainable Development (London: Earthscan).

Moyo, D. (2012) *Winner Takes All* (London: Penguin Books).

NRC (2008) *Minerals, Critical Minerals, and the US Economy*, Report of the Committee on Critical Mineral Impacts on the US Economy, Committee on Earth Sciences, National Research Council (Washington, DC: The National Academies Press).

Norgate T.E. et al. (2007) 'Assessing the Environmental Impact of Metal Production Processes', *Journal of Cleaner Production*, 15. 838–848.

OECD (2011) *OECD Due Diligence Guidance for Responsible Supply Chains of Minerals from Conflict-Affected and High-Risk Areas*, OECD Publishing, Paris. Available at http://www.oecd-ilibrary.org

Östensson, O. (2012) 'The 2008 Commodity Price Boom: Did Speculation Play a Role?', *Mineral Economics*, 25, 1.

Otto, J. (2001) 'Fiscal Decentralization and Mining Taxation', *Report prepared for The World Bank Group Mining Department*. Available at: www.resourcegovernance.org

PwC (2012) *Corporate Income Taxes, Mining Royalties and Other Mining Taxes: Summary of Rates and Rules of Selected Countries* (PricewaterhouseCoopers). Available at www.pwc.co.uk

Radetzki, M. (2008) A *Handbook of Primary Commodities in the Global Economy* (Cambridge: Cambridge University Press).

Radetzki, M., Eggert, R.G., Lagos, G., Lima, M., and Tilton, J.E. (2008) 'The Boom in Mineral Markets: How Long Might It Last?', *Resources Policy*, 33, 3.

Rogers J. (2007) *Hot Commodities* (Chichester, West Sussex: John Wiley & Sons).

Roskill Information Services (2011) *Rare Earths & Yttrium: Market Outlook to 2015*, 14th edn. (London: Roskill Information Services).

Sachs, J.D. and Warner, A.M. (1995) 'Natural Resource Abundance and Economic Growth', *National Bureau of Economics Research Working Paper 5398*.

Solomon, M. (2012) *The Rise of Resource Nationalism* (Cape Town: South African Institute of Mining and Metallurgy).

Strauss, S.D. (1986) *Trouble in the Third Kingdom* (London: Mining Journal Books).

Stürmer, M. '150 Years of Boom and Bust – What Drives Mineral Commodity Prices?', *German Development Institute Discussion Paper 5/2013*, Bonn, Germany. Available at: www.die-gdi.de

Takeuchi, K. (1990) 'Japan's Experience in Linking Foreign Direct Investment and Imports of Minerals', *Resources Policy*, 16, 4.

Tilton, J.E. (1986) 'Beyond Intensity of Use', *Materials & Society*, 10, 3.

Tilton, J.E. (2003) *On Borrowed Time?* (Washington, DC: Resources for the Future).

Tilton, J.E. (2014) 'Cyclical and Secular Determinants of Productivity in the Copper, Aluminium, Iron Ore, and Coal Industries', *Mineral Economics*, 27, 1.

Tilton, J.E., Humphreys, D. and Radetzki, M. (2011) 'Investor Demand and Spot Commodity Prices', *Resources Policy*, 36, 3.

UNCTAD (2009) 'The Financialisation of Commodity Markets', Chapter 2 of *Trade and Development Report*, 2009. 53–84. Available at: http://unctad.org

US Interagency Task Force on Commodity Markets (2008) *Interim Report on Crude Oil* (Washington, DC: US Interagency Task Force on Commodity Markets).

WGC (2013a) *World Gold Council Guidance Note on Non-GAAP Metrics: All-In Sustaining Costs and All-In Costs*. Available at: www.gold.org

WGC (2013b) *Responsible Gold Mining and Value Distribution: A Global Assessment of the Economic Value Created and Distributed by Members of the World Gold Council*. Available at: www.gold.org

Wilson, D. and Purushothaman, R. (2003) 'Dreaming with BRICS: The Path to 2050', *Goldman Sachs Global Economics Paper No: 99*.

Wilson, J.D. (2012) 'Chinese Resource Security Policies and the Restructuring of the Asia-Pacific Iron Ore Market', *Resources Policy*, 37, 3.

WTO (2010) *World Trade Report 2010: Trade in Natural Resources* (Geneva: World Trade Organization).

Data sources for figures and tables

Bank of England, available at: www.bankofengland.co.uk

BP *Statistical Review of World Energy*, various editions available at: www.bp.com/statisticalreview

China Customs, available at: http://english.customs.gov.cn

The Economist, Index of nonferrous metals, available at: www.economist.com

EIA, US Energy Information Administration, available at: www.eia.gov

GFMS, *GFMS Gold Survey*, Thomson Reuters, various editions, available at: https://forms.thomsonreuters.com/gfms/

Humphreys, D (2001) 'Sustainable Development: Can the Mining Industry Afford it?' *Resources Policy*, 27, 1.

Humphreys, D. (2013) 'New Mercantilism: A Perspective on How Politics in Shaping World Metal Supply', *Resources Policy*, 38, 3.

ICMM (2012), *Trends in the Mining and Metals Industry* (International Council on Mining and Metals).

ICMM (2014) *The Role of Mining in National Economies (2nd edition)* (International Council on Mining & Metals).

IMF, International Monetary Fund, *World Economic Outlook* (WEO) Database, available at: www.imf.org

IMF (2014) *World Economic Outlook*, October 2014.

Index Mundi, available at: www.indexmundi.com

Johnson Matthey, *Platinum*, various editions.

LBMA, London Bullion Market Association, available at: www.lbma.org.uk

LME, London Metal Exchange, available at: www.lme.com

LPPM, London Platinum and Palladium Market, available at: www.lppm.com

Maddison, A. (2006) *The Global Economy* (OECD).

Macquarie Capital (2014), *Commodities Comment*, 25 February 2014.

Metallgesellschaft, *Metal Statistics*, various editions.

NBS, National Bureau of Statistics, available at: www.stats.gov.cn/english

PwC (2014) *Mine 2014 – Realigning Expectations,* PricewaterhouseCoopers.

RMD, *Raw Materials Database*, SNL Metals & Mining.

Silver Institute, *World Silver Survey 2014.*

SNL Metals & Mining, *Corporate Exploration Strategies*, various editions. Also, SNL newsletter dated 21 October 2014.

WCA, World Coal Association, available at: www.worldcoal.org

WBMS, *World Bureau of Metal Statistics Yearbook*, various years.

WGC (2013b) *Responsible Gold Mining and Value Distribution: A Global Assessment of the Economic Value Created and Distributed by Members of the World Gold Council*, available at: www.gold.org

WGC (2014) *Responsible Gold Mining and Value Distribution, 2013 Data,* available at: www.gold.org

Wood Mackenzie (2008) (formerly Brook Hunt & Associates) *'Copper Costs: Mines & Projects 2007 edition'*.

WSA, World Steel Association, available at: www.worldsteel.org

UNCTAD, United Nations Conference on Trade and Development, Trust Fund on Iron Ore Information.

USGS, *Minerals Yearbook* and *Mineral Commodity Summaries*, various editions.

USGS (2011), Pui-Kwan Tse, *China's Rare Earth Industry*, USGS Open-file report 2011–1042, available at: http://pubs.usgs.gov/of/2011/1042/of2011-1042.pdf

USGS (2014), *Mineral Commodity Summaries*, US Geological Survey.

Index

Note: Locators followed by f and t refer to figures and tables.

Africa
exploration in, 162f
future potential, 228–9
inward investment, 83–4, 152, 174–5
Africa Mining Vision, 167, 172
Agnelli, Roger, 120, 122
AIM, 130
Albanese, Tom, 120
aluminium and bauxite
China's demand for, 47–50, 49f
China's imports, 68–70
exports from Soviet Union, 20
industry consolidation, 29
prices, 24, 88, 92–4, 93f
world consumption, 210f
Anglo American, 29, 118t, 120, 122, 124t, 125, 178, 227–8
artisanal mining, 9–10, 72, 179
ASX, 129
Australia
effects of boom, 6, 159
inward Chinese investment, 4, 80–2, 84–5, 145–7
iron ore industry, 17–18, 73, 80–2, 85–6, 121, 134, 216–17
mine production, 160–1t
mining's economic contribution,163–5
regulation of foreign investment, 155, 171
super-profits tax, 169

Baia Mare mine, 31
bank involvement in commodities, 104, 107, 214
bank support for GDMs, 118, 120, 126–8, 223
BHP Billiton (BHP)
acquisition of Magma Copper, 25
acquisition of WMC Resources, 119, 120, 133
bid for PCS, 122, 171
bid for Rio Tinto, 121
as GDM, 118t, 120, 124t, 125, 132
merger of BHP and Billiton, 29, 114
post-boom restructuring, 227–8
Bougainville mine, 32
Brazil, 32, 51–2, 64–6, 113, 122, 155, 160–1t, 172–3
Bre-X fraud, 28, 130
bribery, 157, 191
BRICs, 51–2
Bumi, 157–8

Cambior, 30
Canada, 27, 32, 80, 119, 122, 129, 160–1t, 163, 174, 196
capital costs, 20, 66, 68, 97
capital efficiency, 4, 28, 37, 115
capital expenditure, 5, 133–4, 134f, 169
capital markets, 5, 141, 148–9
capital, return on, 19, 22–3, 135f, 172
Carajás mine, 17–18
carbon emissions, 98, 221–2
Carroll, Cynthia, 120
CFTC, 102, 105, 107
Chile, 29, 134, 143, 160–1t, 164–5, 181, 220
China
banking system, 55, 59
CCP, 41, 44, 45, 58, 59, 224
CIC, 85–6, 146, 156
coal industry, 46, 62, 71–5, 72t, 146
construction, 49–50, 55, 94, 211
economic growth, 4, 39–40, 40–7, 51, 53–6
economic imbalances, 56–60, 208–9
economic reforms, 38–40

environmental issues, 57–8, 73–6, 79
export growth, 40–3, 42f
impact on world mineral demand, 47–8, 47t, 90
impact on world mineral prices, 89–90
industrialisation, 46–7
industrial production, 39, 42f, 46, 48, 54
investment (FAI), 43–4, 45f, 208
mineral consumption, 47t, 49f
mineral imports, 63–71, 64t
mineral intensity, 48, 57
mineral production, 62, 72t, 228
mining industry, 71–6, 228
overseas mining investment, 79–87, 82f, 144–7, 155
property market, 54, 56, 59, 209
rare earths, 78–9, 197–201
recovery after GFC, 54–5, 93–4
state-owned enterprises (SOEs), 39, 44, 57, 145–6
steel production and use, 38, 46–50, 47t, 49f, 56–60, 209
strategic stock-pile (SRB), 55, 63
urbanisation, 44–7
WTO, relations with, 4, 6, 40, 41, 56, 79, 197–201
Chinalco/Chalco, 84, 85, 147, 153, 168
China Minmetals (Minmetals), 80, 84, 147, 155
CISA, 56, 81, 112, 209
Cleveland Cliffs, 19, 24
CNMC, 80, 84, 147, 155
coal
China imports of, 64t, 70
largest producers, 160–1t
prices, 92–3, 93f, 109–10
Codelco, 117, 118t, 124t, 143
Cold War, 20, 30
commodity index funds, 102–4, 107
commodity investments, 101–4
commodity markets, 88–99
commodity prices, 100f, *see* mineral prices
communities, 31–2, 35, 87, 167, 179, 224–5, 229
conflict diamonds, 191, 197
conflict minerals, 196–7
Congo DR, 79–80, 84, 150, 155, 157, 164, 171, 173, 178
copper
China's demand for, 47–50, 47t, 49f, 57
China imports, 64t, 66–67
costs of, 24, 96–7, 213–14, 218–19
prices, 21, 21f, 53, 67, 92, 93f, 99, 134
production of, 24, 71–4, 72t, 91, 93, 115, 134, 160–1t, 176, 189t
world consumption, 16–17, 22, 47t, 94, 210f, 213
cost curves, 95–6, 172, 215–16
costs of production, 24, 215, 218–19
critical minerals, 192–3, 194–5
Cuddington, John, 91, 100
CVRD, *see* Vale

Davis, Mick, 114, 228
dematerialisation, 14–16
depletion, *see* resource depletion
Deng Xiaoping, 38, 45, 207
Dodd-Frank Act, 196–7

economic development
mineral dependency, 163, 165, 176
mining's contribution to, 33, 163–7, 164t, 176–7, 180–4, 182t
resource curse, 3, 33, 165–6, 180–1
economies of scale, 24, 72, 126, 213, 217–18
EITI, 191, 194
emerging (market) economies
demand for minerals, 52, 209–14
economic growth of, 50–2
emerging market miners
challenges facing, 153–7
governance, 156–7
growing role of, 141–2, 142f
overseas investment, 144–7
reasons for rise of, 147–50
state involvement, 152–3
stock markets listings, 144–5, 149, 150, 157–8
strengths, 151–3
energy costs, 24, 98, 218, 220

ENRC, 124t, 125, 144, 149, 150–2, 157
environmental impacts of mining, 4, 30–1, 35–6, 65, 73–6, 79, 219–22
European Union (EU)
  concern over mineral supplies, 185–6
  EC critical minerals report, 192–3
  EC Raw Materials Initiative, 191–3
  mining sector, 26, 186–9, 189t
  policies for mining, 191, 194, 201–5, 226
  WTO case vs China, 197–201
Exchange Traded Funds (ETFs), 104, 107
exploration
  activity by commodity, 130–1, 131f, 138
  activity by region, 27, 161–2, 162f
  companies (type of), 10t, 131–2, 132f
  economics of, 128–30, 208
  success, 128, 213, 218–19
export restrictions, 79, 173–4, 199–201

Falconbridge, 80, 119, 123
financialisation of commodities, 105–8
Finland, 26
FQM, 150
France, 26
Freeport-McMoRan, 28, 118t, 120, 124t, 133
futures, 89, 103–4, 105–6

GATT, 41, 199–201
Gensler, Gary, 105
Germany, 26, 193
Glencore, 114, 122, 138, 227
Global Diversified Miners (GDMs), 9, 116, 125–8, 132–3, 227–8, 141, 154, 216
Global Financial Crisis (GFC)
  impact of, 2, 42, 53–4, 92, 123, 132, 174, 223–4
  recovery from, 5, 55–7, 78, 93–4, 133, 208
  response to, 55, 92

gold
  boom of 1980s, 26–7
  China's demand for, 47t, 48–50
  China's production of, 72t, 73–4, 160–1t
  costs, 215
  exploration, 130–1
  grades, 217–19
  mining companies, 27, 120, 125, 136, 138
  prices, 26–28, 92, 93f, 94, 107, 136
  production, 160–1t, 189t
  reserves, 148f
globalisation, 7, 52, 125, 149, 174–5, 223, 225
Global Mining Initiative (GMI), 35–7, 179
Goldman Sachs, 51–2, 102, 107, 126, 212
Gorton, Gary, 103
government policies for mining, 167–74, 183–4, 230
Grupo Mexico, 25, 118t, 124t, 143
Guinea, 84, 147, 163, 164t, 167–8, 171, 178, 180
Guyana, 30, 164t

Heap, Alan, 91, 94–5, 98

ICMM, 9–10, 37, 163–4, 179, 180–1
IEA, 16, 22
IIED, 36
IMF, 9, 51, 80, 106
Inco, 118t, 119–20, 122–3, 143, 154
India, 51–2, 57, 65, 99, 143, 160–1t, 173, 211–12
indigenous peoples, 32, 35, 179, 183, 225
Indonesia
  inward investment, 29, 82, 156
  mine ownership, 170–1
  mine production, 160–1t
  mineral exports, 66, 68–70
  mineral export restrictions,173–4, 204, 223–4
  mining's economic contribution, 163, 164t
  see also Bre-X, Bumi and MINT

investors
  commodity funds, 102–7
  disillusion with mining, 12, 22–4, 28, 37, 133, 136–7, 208, 227
  renewed interest in mining, 2–3, 90, 91, 127–8
iron ore
  Big Three producers, 110–12, 216
  capacity growth, 95, 216–17
  China's demand for, 38, 47t, 48–9, 63–6, 64t, 65f, 94
  costs, 73, 97
  price of, 18, 21, 21f, 92–3, 93f, 111f
  pricing system, 5, 18, 108–13, 125, 133
  production, 72–3, 72t, 134–5, 160–1t, 189t
  reserves, 148f
  slow growth in 1980s, 17–19

Japan
  concern over mineral supplies, 26, 79, 185–6
  iron ore and steel, 17–18, 38, 58, 63–4, 109–10
  mineral procurement policies, 26, 63, 85, 109–10
  rare earths case, 197–201
  slowing growth, 13
Jerrett, Daniel, 91, 100
Jiang Zemin, 39, 41
JOGMEC, 26

Kazakhmys, 144, 149, 157
Kazakhstan, 82, 123, 141, 143–4, 157, 169
Kimberley Process, 191, 197
Kloppers, Marius, 120

labour market shock, 52, 99
legitimacy of mining, 34
Leontieff, Wassily, 16
licence to mine, 34, 179, 229–31
London as mining centre, 3, 8, 25, 138, 214
London Metal Exchange (LME), 12, 17, 88, 214
London Stock Exchange (LSE), 5, 26, 114, 130, 138, 143–4, 149, 150, 157

Mackenzie, Andrew, 137
Marcopper, 31
McKinsey Global Institute, 166, 176, 211
mergers and acquisitions (M&A), 5, 117–23, 127, 133, 136, 138, 154, 228
Metalloinvest, 124t, 125, 144
metal trading, 88–9, 107, 126, 214
Mexico, 5, 52, 143, 200, 212
mineral-consuming countries,174–7, 176f, 185–6
mineral demand
  future drivers, 209–13
  intensity of use, 14, 14f, 39
  shift to emerging economies, 50–2, 209–14, 210f
  slowing in 1980s, 13–16, 22
  see also China
mineral prices
  boom (2003–8), 89–92, 93f
  collapse (2008), 92–3
  declining trend, 20–2, 21f, 28, 98–101, 215
  long term drivers, 215–19
  resurgence (2009), 93–4, 132–3
mineral processing, 172–3
mineral-producing countries, 174–7
mineral reserves, 129–30, 148f
minerals, definition, 9
mining companies
  capital management, 2, 137, 208
  classification, 9–10
  customer base, 2, 7, 108–10, 209, 214–15
  investment cutbacks (2013–14), 136–9, 227
  investment surge (2011–12), 133–4, 134f
  largest, 24, 117, 118t, 124t
  leadership changes, 120, 137
  shareholder base, 127
  sustainable development policies, 179, 180–1, 226

mining industry
  economic contribution, 33, 163–7, 164t, 176–7, 180–4, 182t
  future challenges, 113, 215, 225–31
  market capitalisation, 9, 12, 26, 117, 125, 138–9, 141
  perceptions of, 2, 126, 139, 178
  politicisation, 222–5, 226
  profitability, 3, 22, 23–4, 23f, 28, 135f, 169, 216
  restructuring, 2, 116–25, 227–8
  structure, 9–10, 10t, 118t, 124t
  transformation of, 1–3, 138–9, 229–30
mining investment trends, 94, 115–16, 125–6, 134–5, 134f, 137–8, 202, 207–8
mining waste, 220
MINTs, 212
MMSD, 35–6, 165–6
Mongolia, 82, 146, 147, 164t, 170, 180, 183, 226
Morgan, Hugh, 12, 28

nationalisation, 170–1, 175
nationalism, *see* resource nationalism
National Research Council, 194
new economy, 14–15, 22
nickel
  China's demand for, 47t, 48, 50, 57, 94
  China imports, 64t, 67–8, 70, 77
  costs of, 68, 219
  prices, 21, 21f, 67, 90–4, 93f
  production of, 20, 72f, 160–1f, 176
  world consumption, 14, 16–17, 22, 47t, 94, 210f
nickel pig iron (NPI), 68, 109
non-governmental organisations (NGOs), 31, 35
Noranda, 80, 118t, 119, 120, 147
Norilsk Nickel, 10, 118t, 124t, 140–1, 143–4, 152, 154, 156

OECD, 9, 55, 106, 191, 227
oil and gas, 2, 8–9, 24–5, 100f, 133, 145, 166, 176–7, 176f, 180, 199, 218
oil shocks (of 1970s), 3, 13–14, 16–17, 22
Ok Tedi mine, 31
Omai mine, 30
ore grades, 177, 214, 217–19, 220–1
Oyu Tolgoi, 178, 180, 226

Papua New Guinea (PNG), 31, 32, 82, 164–5, 164t, 206
partnerships, 74, 81, 155, 167, 179–84
'peak' minerals, 218, 222
Peru, 29, 82, 85, 147, 160–1t, 164–5, 169, 183
Phelps Dodge, 24–5, 118t, 119, 120
platinum group metals (PGMs)
  China's demand for, 47t, 48–50, 94
  China imports, 70
  as critical minerals, 171, 192, 194–5
  prices, 93–4, 93f
  production of, 19, 219
  reserves, 62, 148f
  world consumption, 47t
prices, *see* mineral prices
private equity investors, 2, 227–8
privatisation, 5, 80, 142, 144, 147, 148
Prokhorov, Mikhail, 140–1, 144

rare earths, 6, 78–9, 188, 192, 195, 197–201, 198f, 203
rent distribution, 96, 167–9, 183
resource availability, 7, 62, 145, 218–19
resource curse, *see* economic development
resource depletion, 98, 218–19
resource efficiency, 35, 55, 192–3, 213
resource endowment, 3, 4, 24, 62, 81–2, 148, 159–61, 187
resource nationalism
  export controls, 172–4
  local ownership, 170–2
  reasons for, 170–4, 223
  responses to, 87, 152, 155, 177–80, 203–4, 225
  structural drivers, 6, 174–7
  tax increases, 168–9

Rio Tinto (RTZ)
acquisition of Alcan, 120–1
acquisition of BP Minerals, 25
forecasts of demand, 133
as GDM, 118t, 120, 124t, 125, 132, 154
partnering with China, 81, 85, 147, 153
post-boom restructuring, 178, 227–8
Rio 4, 108, 112
RTZ merger with CRA, 29
RTZ as sole LSE listed major, 26
Rogers, Jim, 102
Rouwenhorst, Geert, 103
Russia
exploration, 151, 162t
export surge (1990s), 19–20
export tariffs, 173
inward investment, 29, 171
mine production, 19, 160–1t
mining companies, 123, 140–1, 143–4, 149, 150, 154–5, 156

Sachs, Jeffrey, 33, 165–6
Sierra Leone, 84, 169, 171
Simandou, 84, 153, 167–8, 171, 226
South Africa, 27, 29, 141, 160–1t, 171, 173, 183, 202
Sovereign Wealth Funds (SWFs), 2, 4, 127, 166, 226
Soviet Union, 2, 13, 19, 30, 140–1, 143, 148, 202, 207
speculation, 101–2, 105–7
state capitalism, 175
state-owned companies, 5, 39, 44, 57, 145–6
state's role in mining, 2, 19, 62, 152–3, 170–1, 175, 224, 226
steel
companies, 80–5, 110, 112, 144–5, 149, 154
consumption, 14, 14f, 47t, 210f
production, 17–19, 38, 212, 221
use in China, 46–50, 47t, 49f, 56–60, 209, 211
Strauss, Simon, 7, 206–7

super cycle, 4, 52, 91–101, 116, 119, 123, 126, 129, 133, 135, 169
sustainable development, 34–7, 179–81

tailings, 31
taxes and royalties, 6, 75, 168–9, 177, 180, 182–3, 194
technology bubble, 12, 90
technology metals, 3, 185–6, 213
Teck (Teck Cominco), 29, 86, 119, 124t
terms of trade, 4, 7, 99–100, 215
Tilton, John, 98
TSX, 129

UK, 117, 123, 188, 191
UNCTAD, 106, 172
USA
concern over mineral supplies, 77–8, 185–6
critical mineral studies, 194–5
economy, 13, 48, 52, 90–1, 100, 136, 223
mining companies, 24–5, 48, 117, 207, 226
mining sector, 18–19, 27, 160–1t, 162f, 163, 186–9, 189t, 194
policies for mining, 6, 31, 187–8, 196–7, 201–5, 226
WTO case vs China, 6, 179, 197–201
USBM, 187
US dollar, 10, 100–1, 101f
USGS, 187

Vale (CVRD)
acquisition of Inco, 120, 133, 143, 154
bid for Xstrata, 122, 225
Brazil iron ore, 17, 61, 110, 122, 142, 216, 228
created from CVRD, 142, 143, 149
as GDM, 118t, 124t, 125
value added, 20, 67, 170–2, 192
Vedanta Resources, 133, 143, 149

Warner, Andrew, 33, 165–6
water use, 220
WBCSD, 34, 36

WEF (Davos), 35
weightless economy, 3, 15–16, 22, 187
Wenzhou, 53, 57
Wilson, Sir Robert, 35
WMC Resources (WMC), 12, 118t, 119, 123, 133, 227
World Bank, 9, 56, 74, 194, 204
World Gold Council, 73, 169, 181–2
World Trade Organisation (WTO)
  China's accession, 4, 40–1, 56
  export restrictions, 173, 199–200, 224
  rare earths case, 6, 79, 193, 197–201

Xi Jinping, 57–8, 83, 209
Xstrata, 114, 119, 122, 124t, 125, 132, 138, 227

Zambia, 80, 83–5, 144, 147, 155, 164–5, 169
Zhu Rongji, 4, 39, 45, 64
Zimbabwe, 170
zinc
  consumption, 14, 14f, 16–17, 22, 47t, 49f, 50, 210f
  prices, 21f, 88, 93f, 94
  production, 71, 72t, 189t
  reserves, 148f

Printed and bound by CPI Group (UK) Ltd, Croydon, CR0 4YY